THE SECRETS OF THE SEED
VEGETABLES, FRUITS and NUTS

THE SECRETS OF THE SEED
VEGETABLES, FRUITS and NUTS

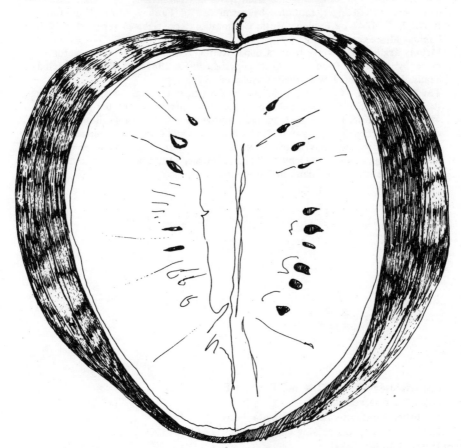

by BARBARA FRIEDLANDER

**Design and Drawings
by IRA FRIEDLANDER**

**GROSSET & DUNLAP
PUBLISHERS, NEW YORK**

ACKNOWLEDGMENTS

Grateful acknowledgment is made to the following publishers, authors, illustrators and other copyright holders, for permission to reprint copyrighted materials.

E. P. Dutton & Co., Inc.—Lines from the poem, "The Little Black Hen," from the book *Now We Are Six* by A. A. Milne. Decorations by E. H. Shepard. Copyright, 1927, by E. P. Dutton & Co., Inc. Renewal, 1955, by A. A. Milne. Published by E. P. Dutton & Co., Inc. and used with their permission, and the permission of the Canadian Publishers, McClelland & Stewart, Ltd., Toronto.

Famous Music Corporation—Line from lyrics of "Tangerine," by Johnny Mercer and Victor Schertzinger, reprinted by permission of Famous Music Corporation, copyright © 1942 by Famous Music Corporation; copyright © renewed 1969 by Famous Music Corporation.

Melody Trails, Inc.—Verse from "Copper Kettle," *The Pale Moonlight,* words and music by Albert F. Beddoe, TRO © copyright 1953, 1961 and 1964, by Melody Trails, Inc., reprinted with their permission.

The New Yorker—Caption for cartoon by Carl Rose on page 74 appeared in the issue of December 8, 1928, and is reprinted with permission of the magazine.

Williamson Music, Inc.—Line from the copyrighted musical composition "A Wonderful Guy," copyright © 1949 by Richard Rodgers and Oscar Hammerstein II; and two lines from the copyrighted musical composition "Oh, What A Beautiful Mornin' " copyright © 1943 by Williamson Music, Inc., copyright renewed; both used by permission of Williamson Music, Inc.

Copyright © 1974 by Ira and Barbara Friedlander
All rights reserved
Published simultaneously in Canada
Library of Congress catalog card number: 72-90854
ISBN 0-448-01368-1
First printing
Printed in the United States of America

To my mother
with deepest love and gratitude

CONTENTS

Introduction xi
VEGETABLES 1
FRUITS AND NUTS 93
  Nuts 152
SEEDS 172
PREPARATION OF VEGETABLES AND FRUITS 175
VEGETABLE GLOSSARY 176
VEGETABLE CHARTS
  Composition of Vegetables 200
  Small-Garden Vegetable Growing Chart 224
  Acid and Alkaline Content 226
  Acid and Alkaline Chart 227
FRUIT GLOSSARY 228
FRUIT CHARTS
  Composition of Fruits 246
  Acid and Alkaline Chart 260
NUT AND SEED GLOSSARY 261
NUT CHARTS
  Composition of Nuts 263
  Acid and Alkaline Chart 266
RECIPES 267

Preface

This is a personal book—eclectic in its gathering of facts, myths, and other thoughts that lie somewhere between the two. I make no claim to exhausting the vastness of the subject. In varying degrees, it covers history, mythology, and nutrition; there are some stories, poetry, and songs; there are charts and recipes. Technical and botanical data have been kept to a minimum. I have drawn on more sources than can be properly credited in this space, and I am deeply indebted to the many friends and passersby who have graciously shared their knowledge with me. I cannot adequately express my thanks to Kathy Komaroff, Marcia Davis, and my mother, Jessie Bankoff, for all their help in researching and compiling; to the editor-in-chief of Grosset & Dunlap, Bob Markel, for his encouragement; to my editor, Kevin Curley, for his invaluable assistance and patience; to Dr. Louis P. Savas for his expert advice; to my husband, Ira, for everything; and to all the plants which provide us with living food and which have inspired this book.

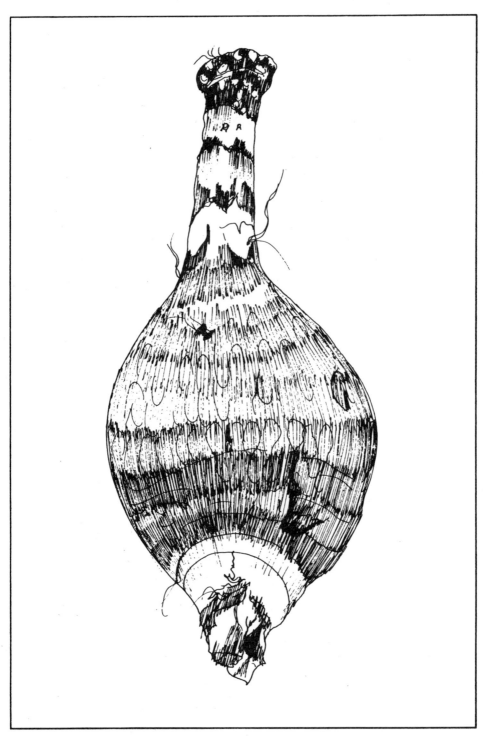

Taro root is a staple in Hawaii where it is called poi.

Introduction

There was no beginning. It was as though seeds had always been there. Man came into a world already equipped with all the gifts of the earth, and he had only to unearth their mystery to have everything he needed. Then, using the creative gift that was his alone, he went beyond his discovery. Perhaps plants could help to explain the mysteries of life and death; wasn't the seed itself a visible example of the cycle of birth, death, and rebirth? He created a mythology and a literature filled with gods masquerading as plants, and vegetables and fruits endowed with human attributes. And so man produced more than agriculture from his garden, for there within the plants were the seeds of religion, dance, drama, medicine, astronomy, etymology, and law.

Sometime around 7000 B.C. man began to tame the wild grasses to make them his cereals. The alleged Garden of Eden was probably situated in the "fertile crescent" of western Asia where the growing plants flourished, showered in winter and warmed in summer. By 3000 B.C. the Mesopotamian region flowered with the oldest of our fruit and vegetable crops, and by the time of ancient Greece and Rome the variety of edible plants was extensive. The Romans expanded cultivation at home and throughout their empire, and it was sustained on feudal and monastery lands during the medieval period of history.

Plants moved slowly from their point of origin to other areas, and so naturally the early varieties of crops were limited in number. But with the age of exploration and the discovery of the New World, the interchange of food, in time, transformed man's diet. Whereas, before, the Romans and Arabs had been responsible for introducing most crops to Europe, now that continent obtained maize, sweet potatoes, and tomatoes from America; apricots and peaches from Asia; and all manner of fruits and vegetables from tropical countries. Meantime, crops from Europe were planted in America, where, in many cases, their production would eventually surpass that in their region of origin. By the eighteenth century, the selection of seeds became an art, and in the nineteenth century, a science. Hybrids were developed by such scientists as Luther Burbank, and manmade chemicals became a part of agriculture. With the Industrial Revolution came mass marketing of food, requiring better means of transportation, refrigeration, and preservation. The earlier methods of pickling, drying, or using heat to preserve fresh food had to be expanded and improved upon. Industry developed processes of canning and freezing, artificial preservatives, dyeing, and waxing. Farmers had to produce more; stronger fertilizers, insecticides, and artificial ripening agents (such as ethylene gas) were required. The story has no end. Today scientists are finding new constructive uses for food plants, but they are also changing the shape, texture, and flavor of vegetables and fruits to conform to increasing mechanization on the farm and in distribution and marketing.

The consumer is the dubious beneficiary of all this, and those who feel that more-natural food is possibly tastier and healthier can still find it, though it does entail more effort. There are alternatives. Certain stores and food cooperatives now carry produce farmed organically using compost, fertilizers without chemicals, and insecticides without poisons*.

Vegetables, Fruits, and Health: Man took slowly to eating fresh produce. As prominent a medical authority as the Greek physician Galen wrote that eating fruit could cause various diseases, and his influence was felt as late as the Middle Ages. The superstitions surrounding, and the reluctance toward eating, fruit and vegetables were probably aggravated by experiences when poisoning occurred from eating unknown herbs; and the incidence of diarrhea (with symptoms similar to cholera) increased during the warm months when fresh fruit was more available—an association that further served to discourage fruit eating.

Urban inhabitants were most affected. Before modern refrigeration, transporting perishable items was risky. As late as 1830, the New York *Mirror* warned: "Fresh fruits should be religiously forbidden to all classes, especially children." Raw food was particularly shunned; in Europe a common practice was to wash it in polluted streams, with the concomitant danger of typhoid, and this was enough to inhibit an already frightened America. On the other hand, from earliest times the plant was medicine. Elizabethan English herbalists, John Gerard (who also was a physician), Nicholas Culpeper, and John Parkinson drew on such ancient Greek and Roman sources as Pliny in compiling works on the medicinal uses of herbs, vegetables, fruits, and nuts. Folk remedies always existed. Early pharmaceutical indices (pharmacopoeia) served to define and describe specific remedies derived from plants; even today, though their chemical derivatives have monopolized the market, botanical drugs in pure form can still be found.

---

* There are some additional methods of farming that go beyond organic growing techniques: namely, biodynamics, which stresses balance, claiming that plants can be either compatible or incompatible with each other and that this factor will influence growth and yield; planting "by the signs," which is as old as astrology and uses the signs of the zodiac and phases of the moon to determine optimum times to plant and to harvest individual crops; and the still-experimental hydroponic method of growing in nutrient-enriched water without soil.

In addition, there are some nutritionists and other individuals involved in natural medicine who advocate using certain fruits and vegetables therapeutically. Despite the range of controversy that seems to exist among food "experts," there is general agreement that it is essential to include fresh fruits and vegetables in a healthy diet. Most vegetables, fruits, and nuts contain at least one and, in many cases, several nutrients, such as carbohydrates, oils, protein, vitamins, and minerals, required by the body.

Fruits and vegetables are our only natural major source of ascorbic acid (vitamin C) and serve to maintain the acid-alkaline balance. (For detailed information, see appropriate charts.)

The medicinal information included in this book is intended to give the reader an idea of how plants have been and are still being used, and no claims of any kind are made or implied.

THE SECRETS OF THE SEED
VEGETABLES, FRUITS and NUTS

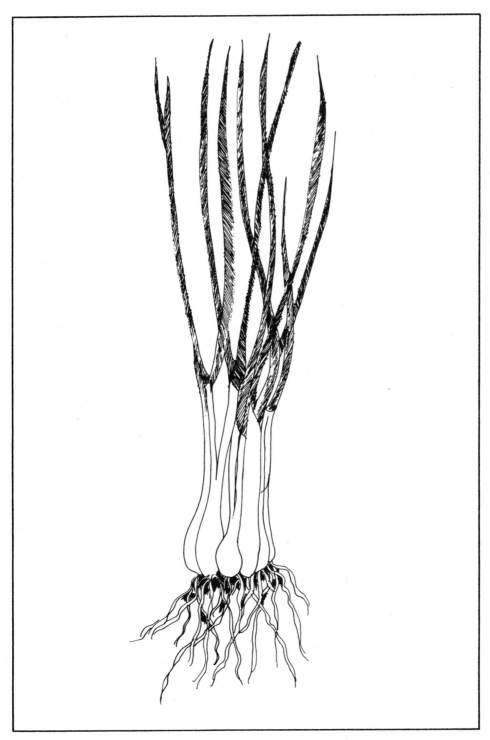

Green onions are often called scallions.

Vegetables

Vegetables can be broadly defined as any edible plant or part of a plant, and it is impossible for this book to list every plant eaten as a vegetable by man. The vegetables discussed here are those that are commonly eaten throughout the world, with particular emphasis on the United States and Europe. Some lesser-known vegetables, some vegetable fruits, and some non-vegetables (like mushrooms) are included. Most vegetables fall into two general categories, leafy and root, and belong to large families bearing even larger Latin names such as Leguminosae (pulses, peas) or Cruciferae (cabbage, turnips). Many of our present-day vegetables are the cultivated descendants of wild plants eaten since prehistory, and even today herbs and vegetables growing wild are eaten. The history of vegetable cultivation is covered briefly in the general introduction. Vegetables had a prominent place in the diet of ancient Rome (meat usually being available only to the very rich), and the Romans contributed at least fifty soups with vegetable bases to early European and British cuisine. Medieval cooking retained this tradition, particularly among the poor. The British took pride in their gardens, and royal gardeners such as Henry VIII's were credited with introducing and improving many strains; during Elizabethan times, vegetables became fashionable.

> "He would live aye; Must eat sallet in May."
> Old English proverb

Salads were eaten at the end of meals by the Romans, but they probably consisted mostly of cooked vegetables. Catherine of Aragon was credited with introducing the use of green leaves in salads into England, although the rich didn't take to them until they became "the thing to eat" during the eighteenth century. Prejudices about raw vegetables took much time to overcome (see introduction), but today salads are not only accepted but enjoyed as a regular part of most diets. Raw vegetables are one of nature's greatest gifts. Chlorophyll, produced by the action of the sun on green leaves, promotes the formation of red blood cells, stimulates breathing, improves circulation, and helps the body to utilize protein. Raw vegetables are particularly effective in restoring the acid-alkaline balance in the body quickly.

The artichoke is related to the sunflower; if left to grow, it produces beautiful flowers.

ARTICHOKES

The word for artichoke in Arabic is *al-khurshuf* or *kharchiof,* which went through many changes to *alcachofa* (Spanish) to *articiocco* (Italian) to *artichaut* (French). It is obvious from the similarities that the vegetable originated in the Mediterranean region. Although this happened thousands of years ago, little mention was made of artichokes until they began to appear as a gourmet vegetable around the middle of the fifteenth century in Italy. Actually, the artichoke is a thistle and a member of the sunflower family and, if not cut down to be used as a vegetable, will yield a beautiful central cluster of violet-blue flowers colorful enough to grace any garden. We eat young flower heads, or chokes; and the heart, which is the fleshy base, is the delicious reward for dipping and stripping the chokes (popularly referred to as leaves). The most popular kind of artichoke is the globe or French artichoke. An earlier type is the cardoon, which is grown mostly for its leaf stalks and eaten more like raw celery than a cooked vegetable. The cardoon looks very much like the globe artichoke, but its leaves are more spiny, its flower heads are spiny-tipped, and it is less appetizing. Cardoons brought a high price in Rome in the second century, and in Spain an extract for curdling milk to produce cheese was made from the dried flowers.

There are two shapes of globe artichokes: conical and globular flower heads. The outside color ranges from light green to purplish green. Artichokes seem to grow best in soil where there is a mixture of saline and alkaline matter, making them somewhat of a maritime plant. They often thrive when the soil is dressed with seaweed, which probably accounts for their high iodine content. Some people consider them an excellent nerve tonic. They require a cool, foggy climate, and in the United States, Castroville in Monterey County, California, is known as the "artichoke capital of the world."

There is some mention of the artichoke's aphrodisiacal qualities, supported by the fact that it was a favorite of Catherine de Médicis, but I have no firsthand corroboration of this. In nineteenth-century English slang, the pun "Have a hearty-choke for breakfast" referred to being hanged.

ASPARAGUS

Asparagus grows wild near so many seacoasts, riverbanks, and lakeshores throughout the world that it's almost impossible to pinpoint its place of origin. We know that the Greeks had a word for it, *aspharagos,* meaning a shoot or sprout, and that it was presented to such notables as Aristotle and Nero by traders from Phoenicia. By 200 B.C., the Romans had begun cultivation, using the wild seeds. The shoots were dried to be used out of the growing season. After Rome fell, the vegetable was cultivated in Syria, Egypt, and Spain. It reappeared in France and England around the beginning of the sixteenth century and was introduced into America in the early 1600s. It has been grown commercially since colonial times, mostly in New Jersey and, later, in California. Wild asparagus looks and tastes like the market variety and can be found in the spring growing beneath the old, dried stalks of the previous year's plants.

Asparagus, a cousin of the orchid, is a member of the lily family, which includes more than 120 species, but the garden asparagus is the only edible species. Some others are used as ornamental foliage, and the asparagus fern is an elegant and popular hanging plant. In the United States we eat only the spears. Fresh green ones are preferred, although white asparagus is often sold in cans. In other areas of the world, seeds are used as a coffee substitute and a spirit is made from the fermented berries. In the Orient, sweetening is common. In Japan asparagus spears are cooked with sugar, and they are candied in China. Asparagus was used medicinally long before it was eaten as a vegetable. The range of ailments treated is long and unrelated; syrups for heart palpitations or as a diuretic; underground stems for bee stings, toothaches, and venereal disease. The actual medical property in asparagus is a substance called rutin, which is a factor in preventing small blood vessels from rupturing. Botanically, asparagus is unusual in that there are distinct male and female plants. The female may be recognized by the appearance of both flowers and seed pods. Mary Washington (a female, naturally) is the name of the most popular strain, but hybrids and an all-male strain are being developed. Asparagus is a perennial that can be productive for up to thirty-five years. Rapidity is a word best used to describe the growth of asparagus; one spear may grow as much as ten inches in a day. Cook as quickly as possible after picking. The Romans had a saying, "Quicker than you can cook asparagus." In some parts, there seems to be an association between the vegetable and a cuckold, probably arising from a Greek myth that likened asparagus to the yield of a ram's horn planted in the ground.

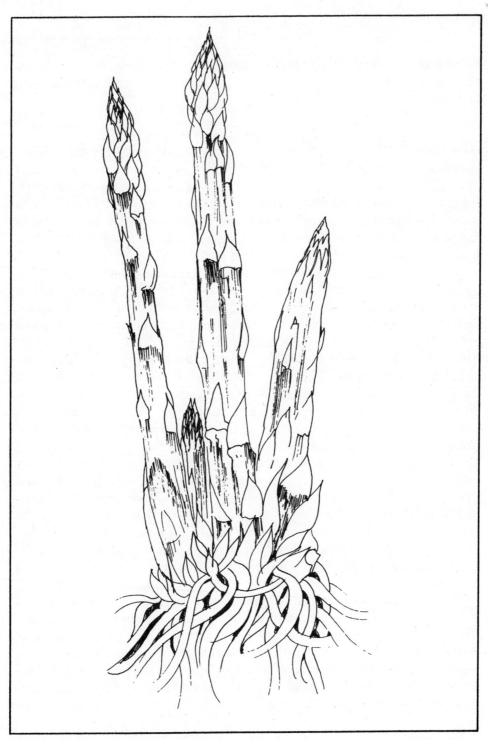
An asparagus spear may grow as much as 10 inches in a day.

BEANS

"And it came to pass, when David was come to Mahanaim . . . and Machir the son of Ammiel. . . . Brought beds, and basins . . . and parched corn, and beans. . . ."

II Samuel 17:27–28

Beans are so ancient that the word often meant any seed, and the original root word was probably associated with the Sanskrit verb "to eat." Beans have been found in Bronze Age sites in Switzerland; Ezekiel was commanded to combine them with grains to make bread; when Daniel was a prisoner in Babylon he refused meat in their favor; they have been found in Egyptian tombs; Homer recorded in the *Iliad,* and Ovid told it too, that bean cakes were offered to the gods and goddesses; in ancient Rome beans were preserved in vinegar and eaten at the beginning of meals (the "dilly bean" is hardly modern!). In Rome also, they had a mystical significance associated with the number nine—in requesting aid for the living from ancestors, the head of the family would throw nine beans over his shoulder nine times. While in India, Alexander the Great found fields of haricot beans. He ordered his cook to prepare them and was so pleased with the result that he brought them back to Europe, where they eventually became the fare of the lower classes. As is sometimes the case with food staples, there also existed a negative attitude toward beans. The Egyptian priests considered them unclean; the Greek and Roman oracles claimed they could cloud one's vision, and consequently upper-class Greeks and Romans would avoid eating them (among Mediterranean people there is a blood disorder called favism, supposedly caused by excessive consumption of broad beans). Literature is filled with superstitions and supernatural happenings connected with beans. The logical Pythagoras believed that, upon leaving the body, certain souls became beans, and because the bean was half-human, he refused to eat it. There is a story that shows how this dogma proved fatal: Pythagoras was pursued by his enemies into a bean field, where he could not bring himself to trample "fellow souls" and, instead, stood still and permitted himself to be killed. Funerals were terminated with beanfeasts; beans were used to exorcize the spirits of wicked souls (in England, several were placed in graves to keep ghosts away, and if you happened to see one, you were to spit a bean at it); the beanstalk was a common Roman funeral plant; Scottish witches rode beanstalks, not broomsticks; roasted beans were buried as a precaution against toothache and smallpox. In Rome, beans were used as ballots both in the courts (white for innocence and black for guilt) and for elections (one of the foremost Roman families took the name of a bean—Fabian) and this is probably the origin of the Latin phrase "abstineto a fabis," which has come to us as "abstain from beans," generally meaning to refrain from involvement in politics (or broadly, as "mixing in"). This custom remained in many European countries until modern times in the election of kings and queens on Twelfth Night and other "beanfeasts." *

Proverbs and expressions with beans in them exemplify the contradictions: "Sleep in a bean field all night if you want to have awful dreams or go crazy" (an old proverb from Leicester, England); "None other lif, sayd he, is worth a Bene" (Chaucer); to "not have a bean" is to be penniless, but to be "full of beans" is to be healthy and high-spirited (possibly because of the energy derived from the high protein content); "not to know beans" is to be appallingly ignorant; and to "spill the beans" is to have a big mouth and upset some plan by talking too much.

"If pale beans bubble for you in an earthenware pot, then you often decline the dinners of sumptuous hosts."

Martial

Beans have been a staple food in many cultures throughout history. They have graced the tables of nobleman and peasant and have fed their livestock as well. (In the United States, broad beans are often called horse beans.) From the time of the Middle Ages there was a death penalty in Britain for the theft of beans, peas, and lentils from open fields.

Beans often come under the broad term of legumes, pulses, or grams (Indian). These words usually refer to those field varieties of beans (many of which are actually peas) that are dried and stored for use in the future, when they can be cooked. They have an extremely high protein content, are rich in nitrates and phosphates, and are a link between vegetables grown under and above the ground.** They can be eaten as "fertilizers" for

---

* Since the time of the Middle Ages, beans were the main ingredient in the Twelfth Night Cake, which also contained honey, flour, ginger, and pepper. This was a sacred cake: one portion was for God, one for the Holy Virgin, and three for the Magi.

** Legumes were treated symbolically as a transitory form between two different aspects of life and played a part in at least one religious ceremony, that of a sect that considered Halloween and the two days following it days of remembrance of the departed and of the saints. For this occasion, a pottage of legumes and rice was prepared. Similarly, "legumes . . . have more animal aspects than any other plant. Their blos-

the body, to tone up the system and create more vitality. (An old remedy for a sore throat was to put white beans in a cloth bag, boil them until soft, and then, while they were still as hot as possible, bind the bag around the neck.) Considering all these factors, it is understandable that beans have been such a popular dietary item throughout the ages and, in fact, have supplanted the more expensive animal proteins in many cultures. (The frequent complaint people seem to have about eating beans is that they tend to be gas-producing. This can be overcome by washing, not soaking, dried beans thoroughly, mixing them with a bit of oil, vinegar if desired, and boiling water, and baking them in a medium oven until tender.)

How does one go about categorizing beans? There are so many kinds and their names change from one country to another; their characteristics differ. The listing below must of necessity be arbitrary and is an attempt at some simplification. Generally, most beans (unless noted otherwise) may be eaten whole, i.e., with their pods, when young and fresh; shelled and dried when mature.

Broad Beans are the beans of history. They are large-podded at maturity and usually shelled. Young pods may be used whole, like French or runner beans. Broad beans are sometimes called fave, horse, or tick. Jack beans, of American origin, are a similar type and are mostly grown in the tropics.

French or Kidney Beans probably originated in South America and have about five hundred varieties. This bean was brought to England and continental Europe in the sixteenth century and soon became so popular there that it replaced many other vegetables and changed eating habits. The British first used the name kidney, probably because of its shape and also to distinguish it from Old World beans. To further the confusion, this bean is also known as the French or haricot bean, whose young pod is usually eaten. (What we commonly call a kidney bean in the United States is actually the seed of the purple climbing, stringless bean.) Garden wax, snap, or stringless beans also belong in this category, although they are called stringbeans by most of us because this is unfortunately a usual characteristic. Some other beans in this vast category, usually used in dried form, are navy (or pea), pinto, marrow, great northern, and yellow eye.

Green Gram (Indian Mung) beans are small green, brown, or mottled seedlike beans, high in vitamin C and enzyme content, which are used mostly

---

soms even look like butterflies held fast. . . . Add to this the very high protein content in the seed which comes from the ability of legumes to assimilate nitrogen from the air. . . ." (Rudolf Hauscka, D.Sc.)

in the Indian dish dhal and for sprouting (see *Bean Sprouts*). Black grams (urd) are a similar variety.

Adzuki or Azuki Beans are small hard, polished beans from Japan. Usually deep red in color, they can also be brown or black. They are always sold in a dried state. A favorite in macrobiotic diets, the adzuki bean is supposed to be very high in vitamin and mineral sources and to contain a substance of benefit to the kidneys.

Soya Beans or Soybeans A Japanese myth relates that the god of the sea, Susano, angrily refused his watery realm and was disowned by the Creator. He wandered around, causing trouble, until he became hungry. Then he visited the goddess of food. But he didn't like what she offered him, so he killed her and crushed her body into the ground. On the site sprang five plants: rice, millet, barley, large beans, and soya beans. Soya is the ancient (since 3000 B.C.) bean of the Orient, and many orientals have lived with this "almost perfect food" as their only protein source. There are over one thousand varieties: white, yellow, brown, black, and multicolored. Sometimes called white gram, these beans, when green, may be eaten like fresh limas; when dried, they may be sprouted or made into a flour (which, because it is low in carbohydrates, is valuable for diabetics and those who are allergic to wheat), soy milk (a useful substitute in diets that preclude cow's milk), bean curd, tofu (fermented soy cheese), and soy oil (the bean is also high in oil). The soya bean was brought to Europe by missionaries during the eighteenth century and did not reach the United States until the nineteenth. In this country it is not so popular as a vegetable and is more of a field crop used for its industrial by-products (although salted, roasted soybeans beat peanuts any day!). It is an excellent soil renovator and livestock food, and entire farms cultivate it for the production of plastics and glue. Fire-fighting foam materials are also made from soybeans.

Lima Beans probably originated in Guatemala, were brought to perfection in Central and South America, and named after the city in Peru where some Spanish explorers found them. (Seeds have been found in mummy pits there and lima plants grow wild in the Amazon basin of Brazil.) The bean traveled up through Mexico and was cultivated by Indians in the southwestern and eastern sections of this country. Since recorded time, limas have been associated in America with corn. In 1605 Champlain reported that the Indians in Maine planted three or four beans with their corn because the beanstalks provided a natural pole and the beans would interlace with the corn as it grew. Soon after the Pilgrims arrived in 1620, Miles Standish dug up an old Indian pit and found some corn and a bag of beans; the Indians had obviously invented the famous dish of succotash—corn and lima beans—years before. Limas are also known as Madagascar, wax, or butter beans. Countless varieties are grown; the best fresh beans are green baby limas that are sufficiently mature to shell from their pods but are not fully ripe. Dried white limas, in all sizes, are more common. Because they were easily stored for long periods and an excellent concentrated food, many explorers and owners of slave ships carried them from the Americas to all parts of the world, where they are grown and eaten today.

Lentils
"And Esau said to Jacob, Feed me, I pray thee, with that same red pottage. . . . Then Jacob gave Esau bread and pottage of lentiles; and he did eat. . . ."

Genesis 25:30, 34

Thus did Esau sell his birthright for a stew of red lentils. The lentil is one of the legumes of antiquity; remains have been found in Stone Age sites in Europe and in Egyptian tombs dating back to 2400 B.C. The Egyptians held lentils in the highest esteem; they were served liberally to children because it was believed that the beans enlightened their minds, opened their hearts, and made them cheerful. On the other hand, the Romans felt lentils slowed the mind. One scholar went so far as to trace the name to the Latin word *lentus,* meaning "slow," and not the more accepted derivation (see below). The lentil is a small, pealike plant (two lentils to a pod) that probably originated in the Near East or Mediterranean region and still has an important dietary place there. It has been traditionally the bean of the lower classes (probably because of cheapness, availability, and high, though incomplete, protein content); in fact, the Hebrew name *adashim* comes from the word meaning "to tend a flock," signifying that lentils were a common dish of the shepherds. Few Indian meals would be complete without a dhal

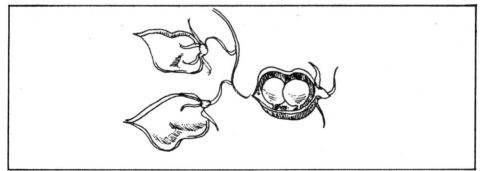

Lentils

(pureed beans) made from red lentils. Lentils are brown, reddish, or orange; the latter have a more subtle flavor and are quicker-cooking. Because lentils split (the name comes from the Latin *lens,* which is also applied to convex shapes such as glasses), they do not require as much cooking time as dried beans. Sometimes the husk of the brown lentil is difficult to digest and can be strained out.

Bean Sprouts   No discussion of beans would be complete without some mention of sprouts. Most dried beans, peas, and seeds can be sprouted to produce greens rich in vitamins, minerals, protein, and enzymes. Soybeans are probably the most familiar sprouts, owing to their prominence in Chinese cuisine; they contain a great deal of protein and some of all the necessary amino acids. One can purchase soybeans cultivated just for sprouting; mung is another bean to sprout. Sprouting beans is a wonderful way to have a kitchen garden year round and provide salad greens when other leaf vegetables may be unavailable. There are many methods; a simple one is to soak the beans overnight in a wide jar covered with a piece of cheesecloth or paper towels. The next day pour off the water, rinse the beans, and place the jar on its side in a dark place. (In a dialect of Chinese, when someone disappears from public he is said to be "sprouting at his home.") Rinse and drain the beans regularly, at least twice a day. Within three or four days the sprouts should be ready to eat; don't let them grow more than one inch tall (or else they will be in a less valuable nutritional state). Sprouts can be stored in a tight container in the refrigerator for up to a week; in fact, unlike other vegetables and fruits, the vitamin C content in soybeans actually increases with storage in the refrigerator. These are delicious and health-giving additions to salads and sandwiches (raw), or to soups and other cooked dishes (cook only a few minutes). Some people claim that they are healing agents for many conditions, particularly kidney and bladder disorders.

BEETS

The beet that we eat as a vegetable is technically a beetroot and closely related to sugar beets and chards. It is also related to the mangel-wurzel or mangold-wurzel (in German *mangelwurzel* means "beetroot"). The mangol beet is used for livestock feeding. The wild beet (*maritima*) probably originated along the seashore of the Mediterranean and still grows in Mediterranean countries, as well as Asia Minor and the Near East. Edible beetroots were relatively unknown before the Christian era. The leaves were eaten, but the root was used medicinally as a remedy for headaches or toothaches or snuffed up the nose to promote sneezing (for the latter, the stronger the scent the better, and the white beet was better). Beetroot was also an excellent tonic for the blood, and for this use baking was preferred over boiling, since the valuable mineral salts are best conserved by that process. By the third century epicures in Rome were creating recipes for beetroots, but history loses the vegetable until the fourteenth century in England, and beet cultivation did not become important in Europe and the United States until around 1800.

The anthocyans in beets influence cell metabolism; highly concentrated powdered beets are being experimented with in cancer treatment.

Edible roots of beet are two colors: red (most used here) and yellow (more common in Europe). Colors can range from dark purplish red to bright red (this type associated with turning "beet red"). Sometimes beets will show light and dark rings when cut open. Red beets will invariably turn the cooking water (as well as one's stools) red. The yellow beet is sweeter-tasting and not used for pickling. The most popular beets are globe-shaped, but spherical ones are also grown.

Beets are easily dehydrated and were used extensively by the military during World War II. The greens are still used in the same way as spinach (see *Spinach*) and have much the same oxalic acid content. Because the garden beet is so closely related to the sugar beet, it is not surprising that its natural sugar content is high, and the beet is often used in producing beer and wine. In England a beet is sometimes added to apples in producing cider to give a more golden color.

At the beginning of the nineteenth century a German scientist discovered that certain yellow-white types of beets, although edible, contained large amounts of natural sugar and that they could be used to produce commercial sugar more easily. This production was expanded during Napoleon's blockade of Europe, which halted the importation of cane sugar. From then on, the industry flourished, because beet sugar was found to be technically superior. It requires about a hundred pounds of root to produce five pounds of sugar. The pollen of sugar beet plants is wind-borne, and fertilization of plants of the same species can take place even over long distances. Garden beets and sugar beets should never be planted near each other because cross-pollination takes place readily and can ruin the seeds of both.

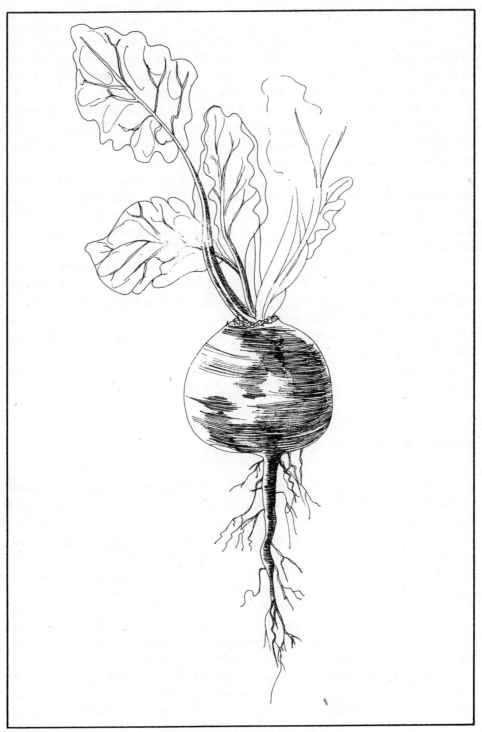
In the U.S., red beets are more common; the leaves may be eaten like spinach.

BROCCOLI

Broccoli and cauliflower are so similar that they are both botanically known as *botrytis,* which means "cluster" in Greek. Broccoli is sometimes called Italian asparagus—a strange conjunction, for although broccoli and cabbage have a common ancestry, asparagus is a relative of neither. The name *broccoli* comes from the Latin *brachium,* meaning "arm" or "branch." There are two general types of broccoli: heading or cauliflower, which has dense white buds like a cauliflower; and sprouting, which has branching clusters of green flower buds on top of a thick green stalk and smaller clusters resembling sprouts along the stem at the leaf attachments. Green sprouting broccoli is sometimes called calabrese. This is the most usual type of the vegetable found in markets in the United States. Sprouting broccoli can also be white or purple, the latter turning green when cooked.

Broccoli was introduced into England around 1720 and has only become popular in the United States fairly recently. Surprisingly, it is an ancient vegetable and was known for more than two thousand years in Europe. It was a native of the Mediterranean area and Asia Minor. In the second century A.D. the Romans grew and revered the same kind of sprouting broccoli we eat today.

The flower buds of the vegetable are so attractive that they are sometimes used in floral decorations.

BRUSSELS SPROUTS

Brussels sprouts are so named because it is supposed that they have grown in the region of Brussels in Belgium since ancient times. Whether or not this is true, it is inaccurate to refer to them as "brussel" sprouts—a frequent misnomer. The vegetable is one of the few to have originated in northern Europe and is descended from the wild cabbage. The Brussels sprout is a relatively "new" vegetable, having been known for only about four hundred years. It was introduced in the United States around 1800 (and grown commercially only since the early 1900s), and it was not popular for most of that time. The Brussels sprout has much to overcome in England, too, for although the English eat it in great abundance, it is usually scorned, probably because it is usually overcooked.

Brussels sprouts grow close together on a tall single stem. The vegetable resembles a miniature cabbage, and its rosette size and shape are so decorative that it is sometimes used in a centerpiece "tree," wired into a pyramid

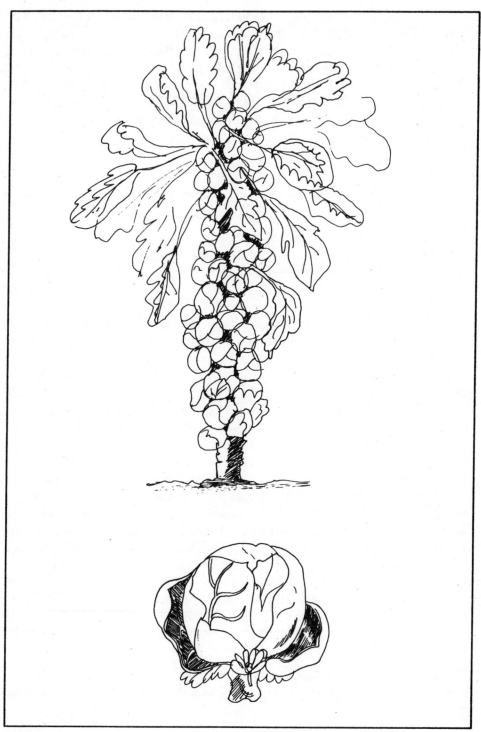

Brussels sprouts look like miniature cabbages.

shape and gilded—beautiful and inedible. The Brussels sprout is an excellent cold-weather vegetable, and light frost seems to improve its taste. In the United States it is grown mostly along the Pacific coast between Monterey and San Francisco and on Long Island. As it is a type of cabbage, the Brussels sprout can hybridize easily with other members of the cabbage family and, when they are growing, care should be taken to keep the plants separate, as cross-pollination can produce unpleasant mixtures.

CABBAGES

> "The time has come," the Walrus said,
> "To talk of many things:
>   Of shoes—and ships—and sealing-wax—
>   Of cabbages—and kings—"
>
> Lewis Carroll

The Greeks told a tale of Lycurgus, who destroyed Dionysus' grape vine. He was punished by being bound to the vine, and his tears took root to become cabbage plants. Another version has them growing from the perspiration dropping from Jove's face. The myth substantiated the belief that the cabbage was the enemy of the grape * and, as such, could cure drunkenness. Aristole and, afterward, prominent Romans left little to chance and, being heavy drinkers, would eat cabbage before going to some wine-laden feast. Cabbage was introduced into Rome several centuries before the time of Christ, and Pliny noted that for about six hundred years thereafter the Romans used no other medicine. Obviously this was a gross exaggeration (the empire would have declined and fallen much earlier), but it is true that cabbage has a high vitamin C content. Poultices made from cabbages, particularly red ones, were used to heal warts, tumors, and sores and were recommended (especially for those born under the sign of Capricorn) as a prevention for rheumatism. Raw sauerkraut juice is still considered excellent for a weak stomach, liver, or bladder. (Cabbage and sauerkraut juice is high in vitamin U, which helps heal stomach ulcers, and in vitamin K.) The cabbage was as revered in Greece and Rome as the onion in Egypt, and so sacred

---

* In the Bio-Dynamic method of farming, some plants such as strawberries are absolutely allergic to cabbage, but aromatics such as sage, rosemary, and peppermint are most compatible.

that one swore oaths on it. But one can overdo a good thing: the Roman consul Cato, in the second century B.C., attributed such mystical powers to cabbages that he believed men could live solely on them as a diet. He remained convinced even after his wife and son died, his cabbages unable to cure their illnesses. Sir Anthony Ashley, who brought "Cato's cabbage" to England, may not have been quite as fanatic but did have a monument in his honor with a cabbage carved on it.

Cabbage is one of the oldest vegetables known to man and has been cultivated for over four thousand years. There seems to be some disagreement about its place of origin: it could have been the eastern Mediterranean region or Asia Minor. Its Latin name *brassica* derives from the Celtic *bresic,* and although the Romans are usually credited with introducing it into Europe, it is possible that the Celts preceded them. The development of the huge cabbage family can be traced in the Mediterranean area and Asia Minor and in northern and western Europe. European names for cabbages (and related vegetables) derive from several sources: Greek *kaulion* and Latin *caulis*, meaning "stem"; Celtic *cap,* meaning "head." The English name is closest to the French *caboche,* earlier *cabus*. The vegetable was introduced into Canada and soon afterward to the United States in the middle of the sixteenth century.

The entire cabbage family includes such seemingly diverse vegetables as cauliflower, Brussels sprouts, broccoli, and kale. Cabbages themselves fall into two general categories: hard-heading and loose-heading. The former was developed in northern Europe and is more adaptable to cooler climates; the latter originated in warmer areas such as Italy and southern France. Both may be either green or white. Round-headed (sometimes called heart or apple) is the oldest hard-headed type; ovoid, flat, pointed, and cone-shaped varieties were later developed. Red cabbage is a favorite with the Dutch and Germans and is excellent when pickled or cooked with apples. Sometimes the red color is superficial, and when one cuts through, there is a decorative design of marbleized red and white. Savoy cabbages can be recognized by their curly leaves. They are slightly milder and grow well in cold weather (September through March). They make excellent coleslaw (a term that derives from the English word for small cabbages, *coleworts*). Less common are flower cabbages, which originated in Japan. These are used in garden borders and for floral arrangements. The inner leaves are bright pink or pale yellow, the outer leaves a variety of shades—all too beautiful to eat!

"The English have three vegetables and two of them are cabbage."
An anonymous gourmet

Cabbage has suffered a poor reputation, largely owing to overcooking, during which its pungent odor heavily permeates the atmosphere. (A Shaker cookbook recommends placing a small piece of charcoal in the water with the boiling cabbage to absorb the smell.) Cabbage need not be cooked for long; in fact, when not eaten raw, it should be steamed quickly in a small amount of water or oil. It will taste far better, and the water should be retained as well, for it contains many valuable vitamins and minerals.

CARROTS

Carrots belong to the same family as celery, parsnips, caraway, and dill and are originally Eurasian. The ancient Greeks called the carrot *philon* from the root word *philo* ("loving"), probably because there was a custom to eat the root as an aphrodisiac before making love. The English word *carrot* comes from the French *carrotte,* which derives from the Latin *carota,* coming from an entirely different Greek word—a verb meaning "to burn." Perhaps the flaming color of the vegetable was the inspiration, as is the term *carrot top* applied to certain redheads.

Carrots grow in all shapes, sizes, and colors. Some Asian types with their bulbous purplish red roots would be unrecognizable to Westerners, who would mistake them for beets. Other colors are pale and deep yellow, red, and even white. Carrot roots may be spherical or cylindrical, and at least one variety in the Far East grows to a yard long. The most popular carrot in the United States is the Mediterranean type, which is fairly long and is deep yellow or orange in color. The best varieties of it have no fibrous central core. Carrots were cultivated in the Mediterranean area before the Christian era but did not become popular as a food until about the thirteenth century in Europe. By the sixteenth century they were being grown all over Europe, and they arrived in America before the Pilgrims. This was one crop the Indians adopted rather than introduced and it soon became a favorite with them. The cultivated carrot is descended from the wild carrot, which is still found growing abundantly. This wild ancestor is actually an herb—with a small, tough taproot that was once slightly poisonous but through cultivation became edible. As with so many other vegetables, carrots were first introduced into man's diet for their supposed medicinal properties. During the first century A.D. Greek physicians prescribed carrots as a stomach tonic; poultices were made from the boiled roots to treat cancers and skin ulcers; the seeds were used to treat snake bite. (Today an extract from carrot seeds is being used experimentally in animals to relax the smooth muscles of the intestines and to depress respiration, which was evidently known by the ancients, who fed carrots to horses to improve their breathing.) Raw carrots are especially rich in carotene, the substance that gives them their characteristic yellow-orange color and that is converted to vitamin A in the liver. For this reason they are supposed to help improve eyesight, particularly night vision. Many health-conscious people take their carrots juiced and find it the most efficacious way to obtain the maximum benefit (an eight-ounce glass provides an average of 50,000 units of vitamin A). A word of warning to extremists: there is possible harm in taking too much vitamin A; and even if you don't overdose, too much carrot juice can turn your skin a bit orange! Because of their natural sugar content, carrots are an excellent sweetener. A carrot cake, which needs far less sugar or honey than other cakes, has a unique and delicious taste; one can even make carrot marmalade flavored with orange and lemon.

Carrot tops (greens) should not be ignored. They are quite ornamental and if left growing in the garden for a second year, will bear delicate white flowers like Queen Anne's lace. The Anglo-Saxons called this a bird's nest, because the tuft of flowers drawn together when the seed is ripe resembles one, and the ladies at the court of King James I often put carrot tops on their hats and in their hair. The greens should be eaten, raw in salads or cooked in soup, because the phosphorus in them is a good source of nerve energy.

CAULIFLOWERS

> "Of all flowers, I like the cauliflower best."
>
> Dr. Samuel Johnson

As previously noted, cauliflower is so closely related to broccoli that the winter variety that matures more slowly is technically referred to as broccoli, but in common usage both summer and winter types are called cauliflower. The vegetable developed from the wild cabbage and came from Syria to the Italian coast and then through Europe. Its French name is *chou-fleur,* literally "cabbage flower," but the English word draws from an earlier source, Latin *caulis,* "stalk." Cauliflower was first brought to England from Cyprus and in the late sixteenth century was referred to as Cyprus colewort (*colewort* is a general term for certain cabbages). The vegetable grows on a single stem, and the head is a crown of undeveloped white flower buds known as the curd. Broad leaves surround the curd and protect it in winter. In hot weather, gardeners sometimes tie the leaves together to guard against too much sun discoloring the white curd. Occasionally one may see what appears to be a purple cauliflower, but then it may be a purple broccoli (see *Broccoli*).

Cauliflowers are sensitive generally to changes in temperature and require a cool moist climate. They should be grown with care. Possibly this is what prompted Mark Twain to make his famous remark "Cauliflower is but a cabbage with a college education." The cauliflower appears versatilely in English slang; in Queen Anne's time it was a comical term for a clerical wig, later referring to anyone who wore powder on his head; sometimes it was the word for the foam on top of beer; and most common is a "cauliflower ear"—mostly found on prizefighters.

CELERIAC (see Celery)

CELERY

The history of celery is difficult to isolate from that of its close relative, parsley. In classical times the two were called by the same name. This is particularly confusing because, reportedly, both were used medicinally. A similarity exists between the leaves of the two vegetables, and these may be used interchangeably. Wild celery (smallage) grew for thousands of years in the Mediterranean region and possibly is the plant *selinon* mentioned by Homer in the *Odyssey*. During the Middle Ages, celery (from the French *céleri*) was employed as a laxative and diuretic; to break up gallstones; to soothe nerves; to heal wounds incurred by animal bites. Even today, celery, which is rich in sodium, is considered very helpful, particularly when juiced with other vegetables, for diseases of chemical imbalance, arthritis, and so on. It was not eaten as a food until the seventeenth century, when the Italians, who are credited with breeding out the bitterness, made it more palatable. By the nineteenth century it was common fare in Europe and the United States. At the beginning of the nineteenth century, someone had the notion that celery had to be blanched of its green color to make it milder-tasting. There are two ways to do this: it can be planted in a trench heaped around with dirt or shielded from the light by wooden boards; or chemically treated with ethylene gas. More recently, "self-blanching" varieties of celery have been developed that are decidedly inferior. All methods of blanching help to destroy chlorophyll, one of the most important qualities of green vegetables. Green celery is available in many areas, and types like pascal are of first quality. In addition to the stalk, which is the most commonly eaten part, celery also yields oil, salt, and seeds.

Celeriac is a turnip-rooted celery that is more popular in Europe than in the United States. It is cooked rather than eaten raw and has a texture similar to that of a potato. It is also often known as celery root or knob.

CHARD

Chard—often called Swiss chard—is actually a white-rooted beet known as a seakale beet,* whose root is not fleshy like other beets. It is cultivated more for its leaves (eaten like spinach) and its stalks (eaten like asparagus). Chard leaves resemble spinach and provide an excellent milder and less acid substitute. Wild chard, like other wild beets, probably originated in the Mediterranean region in prehistoric times. It can still be found there, as well as in Asia Minor and the Near East. The leaves were always used as a potherb, and the cultivated varieties of chard do not differ much from their ancient equivalents. Chard was probably one of the earliest cultivated vegetables. As early as the fourth century B.C., Aristotle mentioned it, but it was categorized with cabbage and therefore considered indigestible and only fit for the lower classes. The Romans ate it too, and it reached China by the seventh century. Times and customs change; by the Middle Ages, soup made with chard leaves was everyday fare. Though not a large commercial crop, chard can be found in markets, in season. Light and dark green are the most common colors, but red chard (resembling rhubarb) is increasingly grown in the United States.

CHICK-PEAS

This ancient legume was probably one of the oldest meals of man, dating from Paleolithic times. It still is prepared in the same form today in parts of Sicily, where chick-peas are cooked in a pan filled with heated stones until tender. The origin of the chick-pea is lost in antiquity. It was, unques-

---

* Please note that the seakale beet is not the same as the seabeet ("wild spinach"), which grows wild on European and Asian shores and can be cooked in the same manner as spinach.

tionably, the esteemed pulse of the Hebrews and Egyptians. To the Egyptians the falcon was the symbolic bird of the principle of self-betterment to which all men should aspire. (Indeed, a book about ancient Egypt is called *Her-Bak* or "Chick-Pea.") If you look at a chick-pea closely, you will notice that it is a miniature falcon's face.

Chick-peas grow in small, hairy pods and are white or cream-colored. Like other peas they may be eaten fresh but generally are dried, roasted, or ground into flour, or soaked and cooked in a variety of ways. Many famous dishes from southern Europe, India, and the Middle East feature chick-peas. The Middle Eastern humus is one of the best known. It is a puree made from mashed chick-peas, sesame paste (tahini), olive oil, lemon, and crushed garlic. Roasted chick-peas are such a popular food in the desert (possibly because they require much chewing and therefore stimulate the flow of saliva) that in Cairo and Damascus there are stands where fried chick-peas are sold to travelers. In Spain the chick-pea is used as a dyeing ingredient for fabrics, as well as a food. The chick-pea's name in Spanish is *garbanzo;* in Italian, *ceci;* in India it is called a Bengal gram; other places, Egyptian pea. The English name derives from a literal translation of the French *pois chiche,* from the Latin name *Cicer,* which also was the root for the famous name Cicero.

CHICORY

Chicory escaped cultivation and was used as a salad vegetable since ancient times throughout the temperate zones. Wild chicory leaves (especially young ones in the spring) resemble dandelion and may be used in the same manner (see *Dandelions*). Chicory is a close relative of the endive, and herein lies much confusion. What we in the United States call endive is actually the leaf of the French *chicorée frisée,* and the curly-leaved chicory is actually an endive. In any case (and by whatever name you choose), the cultivated vegetable is somewhat bitter and its leaves are used mainly in salads, its heart cooked. Some types of chicory are cultivated for their large roots, which are dried, roasted, and ground as an addition to, or substitute for coffee. These roots can also be used in salads. Blanching chicory makes it less bitter, a general rule with similar leaf vegetables, and a very successful commercial blanched type known as witloof or Brussels chicory was developed in an attempt to grow white-leaved vegetables in winter.

Chicory (or succory as it is called sometimes) produces a lovely blue flower, resembling a blue dandelion with petals like the sun's rays. For this reason, probably, it became the theme of one of the numerous sun myths—this one from Romania. There, Florilor, a beautiful lady of the flowers, became the object of the sun god's love. She scorned him because she knew the differences in their positions would preclude a marriage. Angered, he commanded her to become a flower, and she became the chicory whose rays mimic the sun's. It has been called variously sun-follower, bride of the sun, and way-light. Not surprisingly, the seed was used as a love potion.

CHIVES

Chives belong to the lily family and are related to onions. In fact, the name derives from the Old French *cive,* which comes from Latin *cepa,* meaning "onion." A probable native of the Mediterranean region since antiquity, the chive was brought to China about two thousand years ago and was used medicinally (as were most members of the onion family, for they contain an oil that has sulfur in it) as an antiseptic, to stop bleeding, and to counteract poison. The German word for chive is *schnittlauch,* literally, "wound leek."

Chives grow wild throughout the Northern Hemisphere and are cultivated mostly in home gardens. They grow in tufts and do not have swollen bulbs like many other members of the onion family. Only their hollow, thin, grasslike leaves, which grow from three to nine inches long, are used. Chives produce a pink or pale purple flower (the Chinese chive has a white one). They are excellent windowsill vegetables, and if they are repotted and the tops clipped regularly, chives will continue to grow, be tender, and be always ready for chopping into salads, soups, and so on, to impart a more delicate flavor to food than the stronger onion.

COLLARDS

Collards, closely related to kale and sharing the same history, have been traditionally considered the lowest form of cabbage. Today they are much like what they were in their primitive state, about four thousand years ago. Although found growing wild in areas of England and continental Europe, collards (a bastard form of the word *coleworts,* deriving from the Anglo-Saxon; see *Cabbages*), or collard greens as they are sometimes called, are cultivated and have a unique association with the southern part of the

United States. There every farm (and many a yard) has its collard patch. In an earlier time, nutritionists were baffled at the fact that poor southern farmers, particularly blacks, whose diet was sadly deficient and substandard, appeared so healthy and well nourished. This unlikely condition was in a great part due to the heavy comsumption of collards, which have been found to be rich in vitamins and minerals. Today collards with pigs' knuckles or feet are still a favorite "soul food."

CORN

"I tell thee, 'tis a goodlie country, not wanting in victuals. On the banks of those rivers are divers fruits good eat and game aplenty. Besides, the natives in those parts have corne, which yields them bread; and this with little labor and in abundance. 'Tis called in the Spanish tongue 'mahiz.'"
Sir Walter Raleigh

It would be more accurate to call corn *maize,* as everyone outside of the United States does, for the Anglo-Saxon and modern English word corn can mean any kind of grain. A variety of maize (*Zea mays*) is called *Zea saccharata,* and this is the sweet corn we eat as a vegetable on and off the cob. Although one of the grasses that crossed with another to form corn probably dates back to prehistoric Peru, sweet corn did not become a popular vegetable until the mid-nineteenth century. The American Indians who grew it here long before this country was "discovered" did not cultivate the sweet variety, either because they disliked it or because it was more difficult to grow (corn requires careful cultivation). Columbus did take some back to Europe on his first voyage, but it seems that when his sailors tasted it they didn't like it. (A legacy most Europeans still hold to. In Europe cob corn is fed to animals and if an Italian or Spaniard catches anyone eagerly grabbing a few husks along the roadside, he is bound to remark on the crude palate of Americans.) Only 21 percent of the total world supply of corn goes to humans and 6 to 8 percent is used as a vegetable, although it is one of the most important food crops in the United States and a staple in most of South America, parts of eastern Europe, and eastern and southern Africa. Corn is wet-milled for cornstarch, oil, and other by-products; ground into meal and cereal (processed to a fare-thee-well for corn flakes); mashed and distilled for bourbon. Cobs are dried for pipes; and corn is even used to make sizing for magazines. In earlier times the Incas accepted corn as cur-

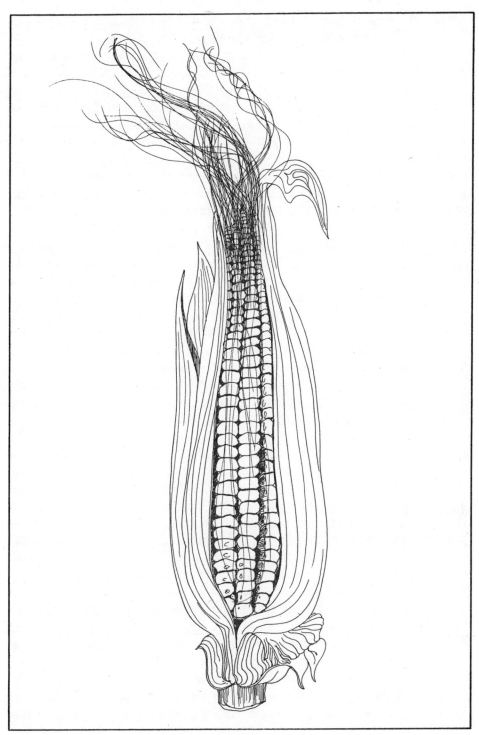

The corn is actually pollinated through the silk; each kernel is a seed.

rency, and even as late as 1670 corn and beans were used for voting on Martha's Vineyard: corn—aye; beans—nay.

Corn (it must be noted that in this sense, *corn* meant any grain) appeared significantly in mythology and folk customs. From the Greek (Demeter and Persephone representing Corn Mother and Daughter) through many cultures up to and including modern Europe, the corn spirit has symbolized the cyclical nature of life itself. The American Indians considered corn a divine gift, and many tribes had legends concerning its origin. Though Uncle Remus may have said, "Crow and corn can't grow in same fiel'," some Indian tribes believed this bird brought the seed from heaven and protected it, so much so that they did not kill crows. One tale, much like that of Apollo and Daphne, tells of a brave who was so in love with a beautiful maiden that he slept outside her hut to offer her his protection. One night, he found her walking in her sleep and followed her. Although she ran fast, finally he caught her, but it was not a young woman whom he embraced. In her fear, she had prayed that she might be transformed and was turned into a tall cornstalk, her hair the silk and her hands the ears. Perhaps the most beautiful symbolic fable is about the sister, White Earth, of the only survivor in the world after all men were destroyed. She was told by her brother that she would be courted by five suitors and to accept none until the fifth appeared. First came Usama. When he was rejected, he became tobacco. Next came Wapako, round and pudgy, and turned into a pumpkin. Third was Ashkossim, a melon, and fourth, Kokees, a bean. Then came a sound like music in the wind, and it was the fifth suitor whom White Earth immediately chose as her husband. The rains came after the wedding and all the previous suitors grew and flourished, but the tallest and best was the corn, her husband, Mondahmin. (This is only a fable, but there are many farmers who will tell you that they can hear corn "grow" on a calm night.) The Indians' reverence for corn was justifiable; it was their dietary mainstay. They ate it roasted right from the stalk (a method the gourmet Nero Wolfe still recommends), combined with beans as succotash, picked green and boiled in some milk pressed from the kernels. They dried old ears and ground them into meal for use during the winter and on long trips. Ground corn was also mixed with water and baked as tortillas on hot stones. The Plains Indians added cornsilk for sweetness and some burned the cob and mixed the ashes with dishes made from corn to preserve the minerals. But the uses of corn went far beyond these. So valuable were its medicinal properties that corn smut or ustilago became official in the American pharmacopoeia as a vasoconstrictor and antihemorrhagic. Dried grains were beaten into a powder for sores and ulcers; fresh ground meal dissolved in water was drunk as prevention for dysentery; a liquid extract from corn silk (the stigmas of the female flowers that receive the pollen from the male tassels) was effective for bladder and urinary dysfunction. Corn oil was used for such diverse ailments as hay fever, migraine headaches, asthma, eczema, and dandruff. One tribe, the Chickasaws, eased itching skin by holding the affected part over the smoke from burning cobs, and the Creeks, to avert some calamity or sickness, fasted, prayed, and ate only a gruel of corn flour and water. If someone died, there was a custom of placing corn on the grave to feed the deceased's soul. Corn contains a protein called zein, of lower nutritional value than other grains, and it has a fairly high starch content. The yellow variety has carotene (converted to vitamin A in the body), but a diet predominant in corn, as is found among some tropical populations and in the southern United States, can cause a deficiency of nicotinic acid (niacin) resulting in the disease pellagra. An interesting fact is that this is not the case in Mexico, although corn is a mainstay in that country, probably because it is cooked in lime water, which supplies the necessary niacin.

> "The corn is as high as an elephant's eye
> And it looks like it's climbing clear up to the sky."
>
> Rodgers and Hammerstein

The Middle West of the United States is the "corn belt" and produces more than one and a half billion bushels per year—more than one quarter of the world supply. (It also accounts for the slang expression "corny," which means something old-fashioned and rural, as opposed to urbane.) It is in this area that the very close relationship between corn and hogs exists, most of the crop being used as feed. In South Dakota, there is an institution called the Corn Palace, where each year a festival is held and corncobs are used as the artist's medium for murals.

Corn comes in many varieties: flour corn, flint corn, popcorn (hard kernels that burst when heated), and, the most popular, dent. Although white and yellow are most common, corn kernels may be purple (almost black), red, and multicolored (often seen at Thanksgiving time and known colloquially as Indian corn). It can vary from four-inch ears on two-and-one-half-feet plants to eight-inch ears on eight-feet plants. Most modern sweet corn comes from hybrid seeds, which the United States developed as a high-yield crop to aid the food supplies of southern Europe after World War II. Corn may be eaten raw and should be cooked only for a few minutes. Anyone who has had it fresh will agree with the farmer's advice: put the water to boil, then go into the fields and pick some.

CUCUMBERS

"Ah, my cucumber, the cucumber gives us a cool mind and we all know the importance of keeping the mind cool. Right from my childhood I've been extremely fond of you and this fondness will forever last."

Sri Chinmoy

It is not unusual to hear an Indian singing the cucumber's praises. It is one of four major vegetables that originated in that country. At least three thousand years old, the cucumber was the first named of the Egyptian vegetables that the Hebrews pined for in the wilderness. " . . . and the children of Israel also wept again. . . . We remember the fish, which we did eat in Egypt freely; the cucumbers . . ." (Numbers 11:4–5). These Middle Eastern cucumbers were cultivated in large open fields in which a crude hut was built for the watchman who protected the plants and guarded them against such marauders as foxes. It was the image of this garden far from the habitations of man that must have inspired Isaiah when he spoke of the desolation of Judah: "The daughter of Zion is left . . . as a lodge in a garden of cucumbers, as a besieged city." There is some mention of the cucumber as an emblem of fertility; one Buddhist legend tells about Sagara's wife who had 60,000 offspring, the first of whom was a cucumber who climbed to heaven on its own vine. Another story, set in ancient Japan, is about a prince who, while walking in his garden, stooped down to pick a particularly luscious cucumber. As he did so, he heard a tinkling voice and spotted a tiny man on one of the plant's leaves who said, "I am the genie of the cucumber and have come to make a bargain. Promise never again to eat one—either you or your family—and I will take you and your descendants under my protection." The prince agreed, placing the cucumber on his crest and growing the vines only for decoration. Though these stories are fictitious, they illustrate the reverence in which the cucumber was held. This carried over into Greece, where one notion was that seeds soaked in cucumber juice before sowing would be protected from insects, and Rome, where the Emperor Tiberius was said to have had cucumbers on his table every day, which probably necessitated the first indoor cultivation by artificial means. The Romans, who seemed to be able to extract that liquid from the most unlikely sources, even made wine from cucumbers. Cucumbers grew in Charlemagne's garden during the ninth century, but they were not cultivated in England until the fifteenth century, at which time they were also brought to America (on one of Columbus's expeditions).

The name derives from the Latin *cucumis*, by way of the medieval French *cocombre*.

As "cool as a cucumber" is an expression based in fact; the interior temperature of the vegetable can be as much as twenty degrees cooler than the outside air on a warm day; its cooling and thirst-quenching properties have been greatly appreciated by everyone, especially inhabitants of warmer climes. It contains more water (distilled by nature) than any other food except its relative the melon—as much as 95.6 percent! For this reason it has often been used as a blood cleanser. In the pharmacopoeia of the eighteenth century, the seeds (which are sometimes pressed to make into cooking oil) were listed as one of the four greater cold seeds and were recommended for their cooling properties, also for coughs. John Gerard thought well of a cucumber stew thickened with oatmeal for people "with flegme and copper-faces, the red and fierie noses (as red as red roses) with pimples, pumples, rubies and such like precious faces." Cucumber juice has been used universally in cosmetics for whitening and softening the skin; a special kind of white cucumber was cultivated in France only for this purpose. When eaten, as well, the cucumber is supposed to help clear the skin, and those English ladies daintily munching on cucumber sandwiches at teatime might well have been giving themselves a facial. Some claim the cucumber will help arthritis sufferers when taken unpeeled and juiced in combination with pineapple juice and a bit of lime (the organic sodium in cucumber is supposed to help dissolve deposits). When possible (which means if you're sure it's not waxed, as many commercially raised ones are) the cucumber's skin should also be eaten. It is suggested that one cut cucumbers starting at the center and avoid eating the extreme ends, possibly because the greatest concentration of toxins would be there. Sometimes running a fork along the skin to make ridges allows for better digestibility. On that subject, there are those who claim the cucumber to be very indigestible, but will acknowledge that the fresher they are, the easier they will be to digest, and that eating them slightly cooked (as is often the custom in the East) makes them more digestible.

Cucumbers may be long or squat and there are even some as round as oranges. They come in a variety of shades of green, or striped, or (rarely) brown. They fall into two general categories, named for the method of growing: indoor and outdoor. Indoor cucumbers are grown with artificial heat and high humidity. Better vegetables are produced if the plants are not fertilized and are therefore seedless (fertilized indoor cucumbers tend to swell at both ends and taste bitter). Greenhouse cucumbers measuring almost two feet have been grown. The most popular type is dark green with smooth skin. Outdoor cucumbers grow on vines like other gourds and must be fertilized. They are usually smaller and have ridged skin. There is a specific species of cucumber called gherkin, which actually is a weed grown in the West Indies—small and very knobby in appearance. What we know as a gherkin (the name comes from the German name for cucumber) is a very small pickled cucumber. The Canadians originally shortened the vegetable's name, and now it is often called a cuke in the United States.

Pickles have possibly been around as long as cucumbers. They were known in the Far and Middle East, and Plutarch wrote that the Romans often used *acetaria* (salads and pickles). Pickling was an excellent means of preserving food long past its growing season and pickled cucumbers had the additional advantage of being considered healthy—particularly for the stomach. Before the beginning of the nineteenth century pickling was done at home in barrels, but since then commercial pickling has become a major industry and certain cucumbers are raised just for that purpose. There are many kinds of pickles, which fall into four main groups: dill, sour, sweet, and fresh-packed. There are also relishes, chowchow, and pickles mixed with other pickled vegetables. To support age-old health claims for pickles comes the evidence from a university study that the vitamin A content of cucumbers might actually be increased during the pickling process. The vitamin C content was long evidenced by the pickle's place in ships' stores as an antiscurvy food; and pickles are supposedly easily digested, owing to the various acids they contain.

DANDELIONS

Yes, the dandelion, that yellow flower growing wild along roads and a "weed" in the grass, is a vegetable. Not the flower, employed more by children and lovers as auguries ("she loves me, she loves me not," etc.), but the leaves (or greens as they are often called), which are used in salads, sometimes raw (quite bitter) or steamed (especially good combined with sorrel). Probably dandelions have been eaten since man first found them. They were the first spring tonic for the colonial settlers after a winter of only root vegetables. In their wild state, they have brought good health to countless people and the ancient herb doctors "prescribed" them often. We now know the reason: dandelions contain as much iron as spinach, much more vitamin C than lettuce, and a large amount of vitamin A. The cooked leaves have been considered an excellent cure for dermatitis and other skin disorders, and the juice is supposed to be very strengthening to the teeth and gums. The roots appear in the pharmacopoeia for kidney and liver disorders as a diuretic. In the Middle Ages, Arabian doctors used dandelion as an eye remedy. Wild dandelion greens should be eaten when they are young and tender, before the plant blooms, in early spring. Only recently cultivated, since the end of the nineteenth century, dandelion greens can be found in some markets in the spring and summer. The cultivated leaves are larger, lighter green, and less bitter. Dandelion is closely related to chicory, and the roots are also roasted, dried, and ground fine to make dandelion coffee, a substitute said to possess all the tonic and stimulant properties of real coffee but without the caffeine. A type of beer is made from the leaves and a wine from crushing the flowers. There is a Ray Bradbury novel entitled *Dandelion Wine,* which describes the golden wine as having the taste of all of summer's warmth to be savored in the winter. The plant is synonymous with summer and bears such names as golden sun and burning fire. It is an emblem of the sun, being governed (in astrology) by Leo. This association may account for its name, which is a corruption of the French *dent-de-lion,* "lion's tooth." Another explanation is the dandelion's toothlike leaf. The plant boasts many colorful names, many coming from children who blew the fuzzy seeds to tell time or fortunes or even the weather: therefore, fairy clock, what's o'clock, weather-glass, etc. More descriptive of its taste are such names as dog's grass, rabbit meat, and stink Davie.

If the need arose, one could make a whole meal of wild dandelion. The roots could be steamed and buttered like parsnips; the crowns (leaf stems on top of the roots) and greens (leaves) could be made into a salad; and you could wash the whole nutritious repast down with dandelion wine and coffee.

EGGPLANTS

Most eggplants are not so ovoid and certainly much too large to warrant the name. But there is a type that grows in the tropics and is white, small, and egg-shaped, and it must have been this that sufficiently impressed the English-speaking people who saw the plants to give them that name. The eggplant is actually a fruit, but eaten like a vegetable. It probably originated in India in prehistory and bears many totally different names in various parts of that continent. The Bengali *brinjal* or *bringall* (corrupted in the West Indies to "brown-jolly") is the most common and even used by the British. On the continent, you will get nowhere by calling it an eggplant;

there it is an aubergine. In some other places it bears such unusual names as Guinea squash and Jew's apple.

The Arabs may have been the first people to think of eating eggplant. They carried it to Spain in the twelfth century, and the Italians popularized it around the thirteenth. But eggplants were not eaten there at that time—only used for ornamentation—possibly because they belong to the same family as the deadly nightshade. (So do the potato and tomato, which, along with the eggplant, are shunned by the followers of a macrobiotic diet.) By the sixteenth century, they were so favored by the Spaniards, who believed them to be aphrodisiacal, that they called them *berengenas* ("apples of love"); northern Europeans of the same period called them mad apples, convinced that they could cause insanity. Some strange legends grew around eggplants. There is an early line of poetry that reads: "Dead-sea fruits, that tempt the eye, / But turn to ashes on the lips," possibly inspired by the purple eggplants that grew along the shores of the Dead Sea and became invested with the terror and superstition surrounding that strange place. It was said that they would be attacked by an insect that would puncture the rind, after which the whole fruit would become gangrenous and be converted to a substance resembling ashes—while the outside remained as beautifully shining as before! One Hebrew historian, Josephus, wrote that they "have a fair color as if they were fit to be eaten, but if you pluck them with your hand, they vanish into smoke and ashes." Even John Milton had something sinister to say about them, calling them "the fruits of Sodom." In time, eggplants became popular everywhere (although only fairly recently in the United States), and in Japan they are considered the third most important vegetable. Large purple ones are most common here, but white and striped ones are also grown. They grow in all shapes (round, oblong) and sizes, and if you can find the miniature ones, cook them whole. They're decorative and succulent. Being in the berry family, eggplants contain many soft small seeds, which are eaten along with the fruit. Eggplant, apart from being a versatile vegetable, is utilized by some Japanese women as a teeth whitener. The custom seems to have come from the tale of a beautiful young woman who wanted to cure her husband of his unfounded jealousy. She dropped eggplant peel into water containing a red solution, turning the water black. When she applied the water to her teeth they became black and ugly. Later, after the desired effect was achieved, she rinsed her mouth and her teeth were whiter than before. One can make or buy a dentifrice containing eggplant. It is made from burned eggplant and sea salt, which is ground into a fine black powder. All you have to do is rub it on the teeth and gums. I guarantee it works and recommend it as long as you don't mind scouring your white bathroom basin each time you brush your teeth!

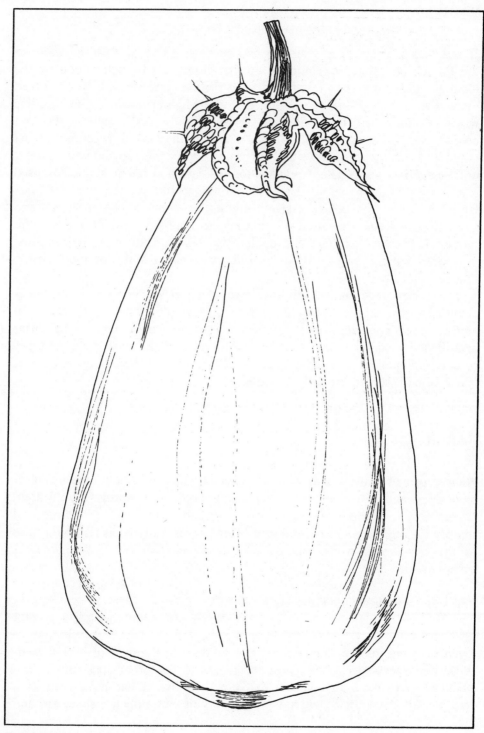

Eggplant

ENDIVES

If one was a purist, one would pronounce the name of this vegetable as the French do (ahn-deev) and, in all likelihood, not be understood by the local grocer. Endive is a relative of chicory; the confusion is already noted under *Chicory*. It probably originated in the Mediterranean region. Endive was eaten long before the time of Christ by the Greeks, and in the first century the Romans used it in salads and as a cooked vegetable. Pliny noted its medicinal properties and suggested endive juice mixed with vinegar and rose oil as an excellent remedy for headache. Ancient magicians used it effectively in their potions. Later the American Indians extracted the juice from the leaves to cool excessive heat in the stomach (heart burn) and liver. Endive seeds were used for fainting, to improve eyesight and remove redness from the eyes, and as a coolant for fevers. Endive is vitamin-filled and, when eaten raw, is still considered an excellent tonic for the bladder, kidneys, and liver.

The varieties of endive fall into two categories: curly-leaved (usually eaten raw in salads) and broad-leaved (an example of this is the Batavian endive, often cooked). As with chicory, endive leaves tend to be bitter and blancing makes them more palatable. One of the most popular types of endive imported into the United States is the so-called Belgian endive, which often is served braised or stuffed.

GARLIC

"Garlic provokes urine and women's courses (menses), helps the biting of mad dogs and other venomous creatures, kills the worms in children, purges the head, helps the lethargy, is a good preservative against and remedy for any plague, sore and foul ulcers, takes away spots and blemishes in the skin, eases pains, ripens and breaks impostumes (abscesses) or other swellings."

Nicholas Culpeper

"... our doctor is a good clove of Garlic."

An English health book, circa 1608

Even allowing for the exaggeration not unusual in the field of herbal medicine, garlic seems to be the oldest and most effective broad-spectrum antibiotic known to man. In addition to Culpeper's list, it has been used as a treatment for malaria (taken with a glass of hot milk), bronchitis and asthma, and the common cold (rubbed on the soles of the feet at night), as a general cleanser of the mucous membranes in the lungs, sinuses, nose, and throat, a preventive for tuberculosis, and for dysentery (by building up the intestinal flora). Externally, it can be used in a poultice for athlete's foot, earache, pimples, lip cancer, and hemorrhoids (by rectal insertion of a peeled clove), and it was often tied around infant's necks to prevent disease generally. And this is probably only a partial listing! Two stories illustrate garlic's effectiveness against the plagues that spread through Europe. During the Great Plague in England in the middle 1600s, while people were dying by the thousands, the occupants of one house in Chester were spared entirely. Upon investigation, it was found that the cellar contained a large supply of garlic. The house was named God's Provident House and remains open to the public. In France in 1721, Marseilles was said to have suffered a plague worse than the one in England. It was difficult to find anyone to bury the thousands who were dying, and the government decided to release four convicts from prison to perform the task. Despite their continuous exposure to the dread disease, all four men remained healthy. The government offered the convicts their freedom in return for the secret of their immunity. They disclosed that every day they drank wine in which much garlic was soaked. This drink became known as the Four Thieves' Vinegar (*Vinaigre des quatre voleurs*), and is still found in France.

Young garlic contains a strongly antiseptic oil called allyl disulphate, which inhibits the growth of bacilli. About twenty years ago a German doctor found garlic oil had amazing antibiotic properties (as the ancients always knew) and began administering it as an actual medicine. Several laboratories in the United States have attempted to synthesize the antibiotic present in garlic and the Russians have already manufactured one, although there does not appear to be much literature about its efficacy. What is known is that there are no side-effects from the habitual or excessive use of garlic. The best way of taking it seems to be chewing raw or inhaling, but garlic capsules (perles), which are odorless, are available in natural-food stores and some pharmacies.

The Chinese were probably the first to cultivate garlic, over six thousand years ago. The Egyptians followed and even employed it in building the great pyramid of Cheops by feeding it daily to thousands of slaves to give them strength and endurance and prevent time lost owing to illness. Egyptians were also allowed to take oaths by placing their hands on a bulb of garlic. Today Egyptians remain among the largest consumers. The ancient Hebrews considered it divine, and Israelite grooms would wear a clove in their buttonholes to signify a happy marriage. The Romans dedicated it to Mars, the god of war; Roman soldiers were fed garlic to give them courage and sexual appetite (fighting cocks are often given it before doing battle). The superstitions and myths about garlic are many: Homer claimed it helped Odysseus escape from Circe (but how?); the Hindus employed it to combat demons, and one name for it in Sanskrit means "slayer of monsters"; it was hung in doorways in Transylvania to keep vampires out and generally hung in houses to drive out the "evil eye"; South American bullfighters have been known to wear a clove around their necks to prevent bulls from charging; and the ancient Greeks put garlic bulbs on piles of stones at crossroads to provide a supper for Hecate, the netherworld's goddess of enchantment.

"Garlic maketh a man winke, drinke and stincke."

Thomas Nashe

"And, most dear actors, eat no onions nor Garlic for we are to utter sweet breath."

Shakespeare

No one is neutral about garlic, and the dissenters choose as their prime target the undeniably strong odor of the vegetable. Enthusiasts have come up with many ways to derive the benefits without suffering the odor. Some claim that it is what remains on the tongue and teeth that causes the odor, and if garlic is minced finely, placed way back on the tongue, and rinsed down with water, no smell will remain. But it has been found that the odor of garlic is breathed into the lungs and so is emitted with each exhalation. There is a story about a famous chef who used to chew a clove, then breathe over the dish he was preparing to give it a delicate flavor. One can cook with a whole head or more and achieve a mild flavor as long as the garlic is not cut—only peeled. The Romans advised planting the bulbs when the moon was below the horizon and gathering them when it was nearest the earth to banish the strong aroma (if planted when the moon is full, the bulb is supposed to become round like an onion, rather than made up of separate cloves). As revered as it was by the labor force in Egypt, the priests considered it so unclean that people who ate it were not allowed to enter the temple of Cybele, the mother goddess of the earth. Alfonso IV, a king of Castile in the middle-fourteenth century, instituted an order of knighthood stating that any knight who had eaten garlic could not appear before him for at least a month. Mohammed said that when Satan left the Garden of Eden, where his left foot fell, garlic grew, and where his right fell, onions. Despite all this and the fact that epicures of old seldom used it in cooking (it was called the physic of the peasants), garlic is a popular ingredient in many dishes and enjoyed all over the world just for the wonderful flavor it imparts. Raw or cooked, it does leave a pungent aftertaste, and the best way to eliminate it is to chew raw parsley, mint, or other fresh green herbs. To remove the odor from hands, sprinkle them with salt and rinse in cold water.

It seems strange that this vegetable, so strongly identified with its odor, is a member of the same family as the onion—the lily family. Garlic bulbs, which grow from individual cloves, not seed, and develop entirely underground, are ready for harvesting when the leaves and flowers that grow above ground wither. Garlic is hardy and can withstand frost. There are various types of garlic, some with as many as twenty to forty cloves to a bulb. The thin paperlike layers range in color from silvery white to almost red.

In some Mediterranean countries, garlic bulbs are strung together and hung.

JERUSALEM ARTICHOKES

If ever a vegetable was misnamed, it's this one, which has no connection at all with Jerusalem and is not even a distant relative of the globe artichoke. The name is an imaginative corruption of the Italian word *girasole,* which means "turning toward the sun," and a more apt name would have been sun-root. This knobby tuber grows like a potato, is a member of the daisy family, and has a flower similar to a sunflower. It is crunchy and sweet-tasting. The Jerusalem artichoke is native American and probably is one of the oldest vegetables grown on this continent. Early French explorers, such as Champlain in Canada, learned to grow and eat them from the Huron Indians and brought the tubers back with them to France in the seventeenth century. A Huron legend explains the vegetable's origin as follows: A Feast of Dreams was held so that the chief's son could choose his life power force. The small boy feasted on a roasted dog and became very sick. In his delirium he called upon the thunder god, who was so pleased that he began to roll his thunder drums, alarming the tribe. The tribe chanted and prayed that another image would reveal itself to the child. Finally the boy's eyes opened and he noticed that sunflowers had grown up all around him. These became his symbol. But the thunder god became angry and sent hailstones down to destroy the herbs. The sun took to battle, pushing the thunder god aside, and caused hailstones to sink into the earth and become edible tubers. These tubers were Jerusalem artichokes, which would provide food and medicine for his people for many years to come.

The Jerusalem artichoke was "discovered" in the Virginia area, where the Indians were found growing them by Sir Walter Raleigh's expedition group in 1585. In Europe, notably France, the vegetable was improved and cultivated, and today there are over two hundred varieties. In England it was used as a sweetmeat in pies. (The sweetness is due to the presence of inulin, a natural sugar that may be eaten by diabetics. This sugar also causes the Jerusalem artichoke to be more difficult to digest than a potato.) Although usually only the tuber is eaten, the leaves may be used as a spinach substitute. A variety known as the American artichoke is ground into a flour from which commercial noodles, breadsticks, etc., are made. Besides the distinctive nutty flavor, these products have the advantage of being excellent dietetic items because the Jerusalem artichoke is 100 percent non-starch.

G. I. Gurdjieff in *Meetings with Remarkable Men* tells of his weakness for Jerusalem artichokes. These artichokes grew tall and thick in Turkestan, providing natural garden and road fences. One day Gurdjieff walked along a road and, feeling hungry, he dug up four large artichokes. He ate three and gave one to his dog, Philos, to try. Philos sniffed it but decided to abstain. After about a week in New Bukhara, Gurdjieff discovered a Jerusalem artichoke on his table. He immediately credited his landlady with great thoughtfulness regarding his love for Jerusalem artichokes. The landlady denied the gesture but the vegetable kept appearing daily. Gurdjieff secreted himself in hopes of solving the enigma and, much to his amazement, he spotted Philos entering his room with a large artichoke in his mouth. Keeping a close watch on the dog, Gurdjieff followed him the next day to the market, where he saw him steal up to a stack of Jerusalem artichokes, snatch one, and depart on a run. When he returned to his room, Gurdjieff found a new artichoke in the usual place.

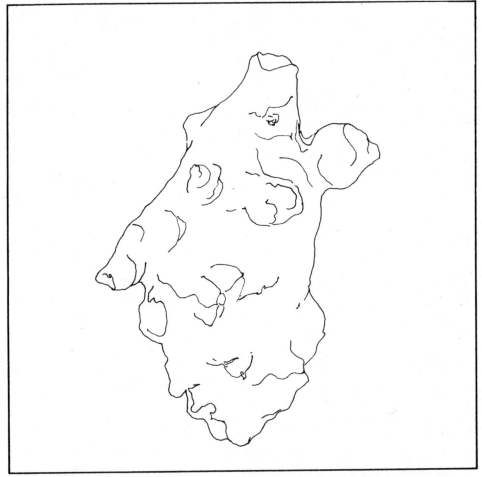

The Jerusalem artichoke's knobby, unattractive appearance belies its sweet taste and versatility.

KALE

Kale is a primitive type of cabbage popular in colder regions as a winter green; its flavor is actually enhanced by a touch of frost. It is a strong-tasting vegetable that is usually cooked, although some "hardies" eat it raw. Kale (and its close relative, collards) closely resembles the original cabbage—leafy and non-heading—and still grows wild in northern Europe and England. The Greeks and Romans cultivated and ate kale (the word comes from the Germanic root *kohl,* occurring also in *kohlrabi, colewort,* and others), which is also known as borecole, or curly greens. Kale is not a large commercial crop in the United States and elsewhere is grown mainly as sheep feed, but it was so important at one time in Scotland that kale soup was the mainstay of most meals there. In fact, not only the soup, but also the pot in which it was cooked, was referred to as kail. Kale is a popular home-garden vegetable and extremely easy to grow. Curly kale is most popular. It looks like large parsley, but it can also be yellow-green, red, purple, and even silver (mostly cultivated for floral arrangements) and have plain or deeply frilled leaves.

Seakale is distinct from kale and is not even a member of the cabbage family. This wild vegetable is probably related to the cresses and grows along seashores in western Europe and the British Isles. It is used as a potherb and sometimes as a substitute for cabbage by fishermen's wives during winter. It has some supposed value in strengthening the urethral ducts.

KOHLRABI

Kohlrabi is that strange, octopuslike vegetable one sees occasionally in the market and rarely buys. The name is German; *kohl* means "cabbage" and *rabi* means "turnip." It is as accurate a name as is possible, for this vegetable is a form of the same species to which cabbage belongs (its leaves may be eaten in the same manner), but the part generally eaten is the turniplike globe. Kohlrabi is a close relative of the Brussels sprout, which appears farfetched until one looks at the leaves of both plants and notes the strong family resemblance. They are two of the newest vegetables existent, since the plants were not used as a food until about five hundred years ago. They are also the only common vegetables (with the possible exception of rutabagas) that originated in northern Europe. Kohlrabi is not a popular vegetable in the United States—its taste can be fairly strong—but if it is picked before maturity, when it is small, it is milder and sweeter than a turnip. It grows easily and has the distinct advantage of surviving both frost and drought. White (more like light green) and purple kohlrabi are grown, but the former is the type most often eaten.

LEEKS

Leeks are the sweet cousins of onions and are so ancient that one can only guess at their place of origin—probably the eastern Mediterranean or the Near East. The leek has been cultivated for from three to four thousand years and is still found growing in its wild form. It was a sacred plant to the ancient Egyptians and is featured on an inscription on one of the pyramids. Like the onion, the leek was seen by the Egyptians as representative of the universe; each layer corresponding to the successive layers of heaven and hell in their cosmogony. Along with the cucumber, onion, and melon, it was one of the foods of Egypt that the Hebrews longed for in the wilderness. To the Romans, the leek was a symbol of virtue because Apollo's mother, Latona, supposedly longed for them. The emperor Nero was so fond of leeks that he ate them with oil (abstaining from bread) for several days each month, claiming they made his voice clear. For this habit, he was given the dubious nickname of Porrophagus, "leek-eater." In A.D. 640, there was a battle between the Welsh and the Saxons. Saint David advised the Welsh to pick some leeks from a nearby garden and wear them in their caps ("the fairest emblym that is worne") so that they would be able to identify each other. The Welsh won a great victory and ever since have worn leeks in their caps on Saint David's Day, March 1.

The word leek is pure Anglo-Saxon. One scholar goes as far as saying it once meant any vegetable; later, any bulbous one (the suffix *leac* appears in many Old English names for various members of the onion family). The word *porridge* used to mean a thick vegetable soup primarily made of leeks. Leeks evidently were so important in England that a *leac-ton* (leek garden) was the common name for any vegetable garden and a *leac-ward* was a gardener. "To eat the leek" is akin to eating humble pie, possibly because of the vegetable's association with the diet of the common people or possibly from Shakespeare's *King Henry V*, where Pistol is forced to eat a leek after he has heartily disdained it.

> "Now leeks are in season, for pottage full good,
> That spareth the milch cow and purgeth the blood."
>
> Advice from an old herbal
for the month of March

Pliny wrote that the leek was used in no less than thirty-two remedies. It was used as an expectorant (Nero may have been right; an old saying goes: "Lovers live by love as larks by leeks"), for clearing the kidneys, for chilblains, chapped hands, and sore eyes. In cooking it has always been best known for its place in the souppot: the pungent Scotch leek is the main ingredient in the famous cock-a-leekie soup and there is no true vichyssoise without leeks. But leeks are also delicious by themselves or as a sweet substitute for onions in many dishes. Leeks have flat leaves, which are folded sharply lengthwise, ending in a cylindrical white bulb. The lower section of these leaves is usually blanced, and most people eat only the blanced part and discard the upper green section. The leaves may be sliced lengthwise and then chopped, if necessary. Leaves and bulbs should be eaten. When preparing leeks, make sure to separate the leaves and wash them very well because sand often clings inside the sharp creases. Leeks too old to eat are easily recognized, because the inner part of the leaves becomes woody and too hard to even cut with a knife.

LETTUCE

By far the most popular salad plant, lettuce has been cultivated for so long that there is doubt concerning its exact place of origin, probably the Mediterranean or the Near East. (In the Jewish feast of Passover, lettuce was probably used as one of the bitter herbs.) Early lettuces, like cabbage, were loose-leaved; firm and loose-heading varieties were a later development. The Persian kings ate lettuce 2,500 years ago, and the Romans had as many as a dozen varieties. The popular romaine lettuce is a long-leaved cos type that gets its name from "Roman." Lettuce became associated in the Near East with the cult of Ishtar and Thammuz (Venus and Adonis to the Greeks) because the latter was said to have been placed on a bed of lettuce when he died. Fertility festivals in Greece in the seventh century B.C. were known as Adonis festivals, and quick-growing lettuce was put in pots and carried to symbolize the transitory quality of life. These potted plants were called the gardens of Adonis and possibly were the beginning of the custom of raising plants in pots around the house. Cultivated lettuce was introduced into England at the beginning of the sixteenth century. King Henry VIII gave a special award to his chief gardener for contributing "a salad of lettuce and cherries."

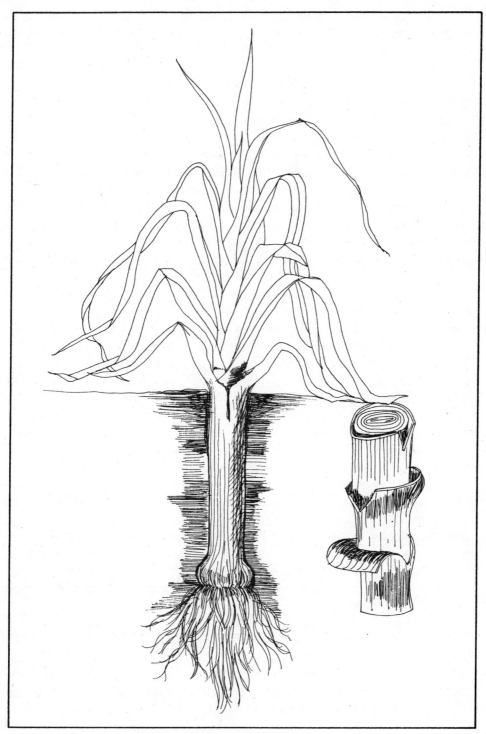

The lower portion of the leek bulb is generally blanched.

"Lettuce seeds brused and mixed with the white of an Egge applyed in plaister form on the temples or forehead warme at the going to rest dooth marvelously procure sleepe."

An old English herbal

Edible lettuce is related to poison lettuce, which is a common European plant. Most lettuce is potentially narcotic, and for this reason was known as *sleepwort* to the Anglo-Saxons. The dried juice from wild lettuce was used as a sedative, hypnotic, and cough-suppressor. Possibly lettuce does contain some laudanum, a soporific. When the Romans ate it, they would "correct" its effect by drinking half a large snail shell of mixed spice, vinegar, and honey. Often it was boiled rather than eaten raw. John Gerard advised, "Lettuce should be boyled in order that it may be sooner digested and nourish more." The name for lettuce comes from this soporific substance, which is the milky juice in the leaves. Latin *lactua* becomes Italian *lattuga,* Spanish *lachuca,* and Old French *laittues*—all similar to words for milk. There was a superstition that for lettuce to be beneficial in healing it had to be pulled up by the root with the left hand before the sun rose and laid under the invalid's pillow so that the larger end of the stalk and leaf lay toward the feet. Lettuce will keep fresh longer if roots are left on the plant, and freshness can be a problem with lettuce as, possibly, it suffers greater nutritional losses from being shipped and stored. As with so many other vegetables, the ideal is to eat it fresh-picked (some varieties can be grown in home gardens), and it must be crisp when used in a salad. To achieve this, some people place lettuce leaves in the freezer for a few minutes before serving (a few minutes is the maximum—lettuce should not be frozen). Soaking in water may crisp leaves, but it also depletes them of their vitamins and minerals, which end up in the water. Insecticides are a particular hazard on lettuce. It is difficult to wash the oily residue off curly leaves (see page 175 for procedures for washing fruits and vegetables). Although leafy vegetables—especially lettuce—are a good cooling staple in warm weather, some "experts" advise care in eating to excess; the nitrates in vegetables like lettuce become nitrites under the hot sun, and these nitrites could unite with the red cells in the blood that carry oxygen, becoming harmful in large amounts.

Many hundreds of varieties of lettuce are cultivated. The two general categories are head (cabbage) and loose-leaf (cos); lettuce can also be classified as summer or winter according to the growing season. In the United States the most popular types are iceberg (more accurately, imperial), which is a hardy strain of head lettuce mostly grown in the Southwest; Boston (loose-leaved, light green, and very delicate); red-tipped; and the previously noted romaine, which has a strong molecular structure that withstands the abuse of shipping. Many consider romaine (and the other loose-leaved types) to be more nutritious than head lettuce, possibly because each leaf has more access to the sun, producing more chlorophyll and becoming greener. Edible wild lettuce, such as Canada wild lettuce, is still found growing abundantly and can be eaten when young.

In 1938 a missionary from western China, close to the Tibetan border, sent a few seeds of a vegetable eaten there to an American seed company. This became celtuce, which looks like a cross between celery and lettuce. It has fleshy stems (whose outer covering is peeled and then cooked), leaves, and a heart that are eaten like letture. Celtuce is an easy crop for home gardens.

MUSHROOMS

"Keep your grain, O Libya, unyoke your oxen, provided that you send us mushrooms."

Juvenal

Mushrooms have probably always intrigued man. The Egyptian pharaohs three thousand years ago forbade any commoner from touching them and thereby ensured their own supply. The Romans monopolized them, too, as food for the aristocracy. Nero called them "food for the gods"—a double-edged phrase, for though mushrooms were prized for their incomparable delicate flavor, the poisonous varieties were also responsible for the deaths of many kings and emperors. The Roman philosopher Seneca bluntly named them "voluptuous poison," but this did not deter many of his wealthy compatriots who ate field mushrooms, uncertain of the species, and felt it well worth the risk. The foregoing refers to field mushrooms; cultivation wasn't even attempted until the time of Louis XIV in caves outside of Paris. England followed, but commercial cultivation did not begin in the United States until the late nineteenth century. Today mushrooms are widely cultivated, but usually it is just one species of field mushroom (*Agaricus bisporus*) that is the main commercial mushroom of Europe and the United States, although hundreds of species are edible. Possibly this can be explained simply by describing the method by which mushrooms are grown.

Mushrooms are fungi, not vegetables at all. They do not contain chlorophyll, so instead of converting sunlight into food they must receive their nourishment from other organic matter like scavengers or parasites. Part of the vast family of fungi that includes rust, molds, mildew, and dry rot,* mushrooms attach themselves to any organic matter, especially trees, which through photosynthesis are able to provide the necessary food for the fungi. A mushroom is actually the fruiting body of the fungus, which remains underground. Mushrooms, unlike plants, do not produce seeds and have no flowers or actual roots. Instead they produce spores (sometimes as many as half a billion) within a fine dust that floats on the air or water and settles in a favorable place for growth. There it soaks up water until it expands and cracks and sends out spawn, a stringy white material, like roots, which attaches itself to the organic matter and in time becomes a mushroom. And then the cycle is completed; the gills beneath the mushroom's head contain the new spores, which, in time, may blow away to begin the process of further reproduction. To note how spores travel easily, place a dark-spored fresh mushroom cap (with the stem removed) on a piece of white paper and cover it with a bowl overnight. The next morning there will probably be a definite print on the paper where the spores have fallen. The reproductive process is most primitive but very difficult to simulate when cultivating mushrooms; so "simple" in nature, it is highly complex under artificial conditions where the temperature and light must be carefully controlled. Also, mushrooms are prey to many diseases. Special windowless houses are constructed for growing mushrooms. One large farm near Pittsburgh, which grows more than ten million pounds annually, is an underground mine; the pickers wear miners' caps to provide light for working. Modern agriculture has developed cultivated spawn that can be inoculated into new beds, but its use is limited to one species which is most adaptable.

Mushroom study (mycology) is a field in itself, and many books have been devoted entirely to it. By necessity, we can only have a cursory look at it here. Field or wild mushrooms appear in as many glorious colors, sizes, and shapes as nature seems capable of, and ironically some of the most beautiful are the most poisonous. A word of caution is in order here: there

---

* Molds, etc., have always been a menace to man. They can attack animal tissue, plants, grain, timber, leather, fabrics, even insulation on wires and coating on optical lenses; in fact, almost any organic material. A fungus blight was responsible for the destruction of potato crops that caused the famine in Ireland during the 1840s. Conversely, molds can be very useful; one bread mold produces penicillin, others sauerkraut, vinegar, and such delicacies as camembert and roquefort cheeses. Many lichens provide the base material for certain dyes and perfumes and provide food for arctic reindeer. Mushrooms perform a necessary task, removing the decay of vegetation, and even poisonous mushrooms feed upon the roots of trees in a valuable symbiotic relationship.

is no foolproof way to know whether a mushroom is poisonous or not—except through the knowledge of an expert mycologist. Before any experimentation, the reader is strongly advised to learn to identify mushrooms through a reliable field guide such as the Department of Agriculture circular entitled *Some Common Mushrooms and How to Know Them*. Also, it is important to learn the Latin names of the more common ones, as local names differ. Poisonous mushrooms are generally called toadstools, although nobody seems to know why; possibly because of the shape. Those belonging to the species bearing the name *Amanita* (usually followed by a more specific name such as *virosa*) are the most poisonous and some, such as the nicknamed Destroying Angel, will cause violent intestinal seizures and usually death within a few days. Generally toadstools are dangerous in proportion to the amount eaten. And because a spore travels easily, be careful not to place a dangerous type near edible ones, for they may become contaminated. Some mushrooms have intoxicating, "consciousness-expanding" properties and have been used by Mexican Indians, Vikings, and others for this purpose. It must be noted that with mushrooms there is a fine line between intoxication and death. In addition, all vision-producing mushrooms may also cause problems to people with allergies or when combined with alcohol or when eaten raw or when they begin to decompose. The solution seems to be: when in doubt, don't eat it, or, if you should feel sick after eating any mushroom, force yourself to vomit and then drink a lot of water. Now, with all the cautions behind us, here is a partial listing of some edible field mushrooms. Many of these mushrooms, particularly those which, reportedly, are the most delicious, such as the boletus, are impossible to cultivate because they cannot survive without being attached to roots of certain plants and trees.

Morels Usually found in the spring, morels are among the tastiest of edible mushrooms. They have a distinctive appearance—spongelike, their caps crisscrossed with irregular pale-brown ridges surrounding darker brown hollows, which house the spores. Morels are seen in dried form all over the world. A favorite in French cuisine, the most famous morel is known in France as *morille délicieuse* and in Italy as *morchella deliciosa*.

Chanterelles Usually found in summer, chanterelles are another favorite field mushroom, particularly in the Alps and south Germany, where they are called *pfefferlingen;* in France they are called *girolles*.

Puffballs Puffballs have white flesh and are so called because a slight touch can open the skin and puffs of spores will blow into the air. Some giant puffballs measure more than three feet in diameter.

Boletus The boletus is porous, with a spongy mass of tubes instead of gills under the cap. The *cèpe,* which is a famous mushroom on the Continent, is a type of edible boletus.

Beefsteaks These mushrooms are usually found on dead logs and are aptly named, for they are the closest to meat of any vegetable. They are juicy and fleshy, dark red above and yellowish underneath.

Truffles Truffles belong in a category by themselves. Their spores do not develop within them; instead, they are carried by insects attracted by the aroma to nearby trees. Truffles cannot be cultivated. They grow entirely underground and absorb the nutrients in the soil to such an extent that sometimes one can identify where they are growing by the barren look of the landscape—where only the trees can grow. More often, they have to be hunted by trained dogs or hogs (the former are preferred since they will obey commands and not eat the find). Truffle dogs are highly prized and very expensive, although often, when one has started digging, the scent of the truffle is so strong that even the human hunter can trace it. Truffles are one of the most expensive of man's food. The two most cherished kinds are the Italian white and the French black of Périgord, named for the town of Périgeux where the best are grown. White truffles are so valuable that their location is a closely guarded secret; digging is often done after dark, and maps of areas where they grow are passed down from generation to generation as a legacy. White truffles are rarely cooked (some are canned, but not worthy of their exorbitant price) but are usually served raw, thinly sliced, just for the unique, slightly garlicky aroma they impart to any food they touch. The black truffle is very black, with a warty exterior. The flesh inside is blackish gray with white veins. It is more plentiful than the white and is usually cooked, although it may be eaten raw. It is also more renowned for its aroma than its almost nonexistent taste. Black truffles are perhaps most famous for their place in pâté. Again, canned are far inferior to fresh.

Miscellaneous Other edible mushrooms include shaggy manes, oyster, and elm. Oriental shiitake and matsutake are large black mushrooms usually found in dried form.

A fascinating phenomena of mushroom-growing are "fairy rings"; sometimes as many as a hundred mushrooms seen growing in a circle around grass in a meadow. Often the grass in the center is withered from the underground fungi. This gave rise to many legends: fairies and elves danced on the grass at night, trampling it and then sitting down to rest on the mushrooms; sheep supposedly refused to eat the grass within the rings, and young women left it undisturbed for fear the fairies might destroy their beauty in revenge. The rings were often considered sacred; a Scottish saying is "He wha cleans the fairy ring an easy deeth shall dee." Some fairy rings can be as much as fifty feet in diameter and disappear in a few years, widening outward as new spores take root. Some, like those in Colorado, are said to be four to six hundred years old, and some on Salisbury Plain near Stonehenge, in England, are of indeterminate age.

There seems to be divergent opinions regarding the nutritional value of mushrooms. There are claims that they are rich in folic acid and vitamin B_{12}, two elements that combat pernicious anemia, and in vitamin D (particularly strange—as this is the "sunshine" vitamin and mushrooms rarely see the sun while growing). Because of their fleshy texture, mushrooms are an excellent meat substitute. The negative position claims that their nutritive value is light and that they may tend to hinder the assimilation of other foods because they are indigestible. Indian Brahmans were prohibited from eating them, as were persons following a yogic diet; the reason for this may be that their classification as fungi and the manner in which they grow are possibly associated with decay. There seems to be no question, though, of their place in good cooking. Both white and brown cultivated mushrooms are good; closed caps usually indicate freshness, and as the mushroom ages, the gills are more revealed. Again, this is a matter of taste; Europeans say open mushrooms taste better. Italian cooks insist that marjoram must accompany mushrooms and call it "the mushroom herb." There are a great variety of dried mushrooms—Italian, French, and oriental—usually requiring some soaking before cooking. And don't ignore the use of fresh mushrooms raw in salads or marinated in appetizers. In French cuisine, when a recipe includes the words *bonne femme,* it means mushrooms are one of the main ingredients.

OKRA

Okra originated in Africa, probably in the region of Ethiopia, where it still grows wild, and it was cultivated in ancient Egypt. When one learns that this vegetable is a member of the cotton family it is not surprising that it is grown abundantly in cotton-raising regions such as Egypt and the southern United States. Okra was probably brought from North Africa to the European continent by the Moors, and it is likely that the French introduced it into this country (early in the eighteenth century), because it is a popular ingredient in the famed French cuisine of New Orleans. Okra came to this country with one of its African names, kingumbo, and is still often called gumbo, although it is more accurate to refer to the dishes that it flavors so well by that name. It is not a very popular vegetable and rarely seen fresh in markets outside of the South, although it is sold frozen or canned. The pod, which contains small seeds, is used whole, sometimes coated with cornmeal and deep-fried or more often, because of its mucilaginous nature, as a thickener in soups and stews (like chicken gumbo). In some countries the ripe seeds are pressed to make into a cooking oil. Dried pods are used as a spice.

Roselle is a close relative of okra. It is little known in the United States. Its leaves are used to make a sour-tasting drink and for jellies and sauces. In Turkey the leaves are made into a preparation to soothe and reduce inflammations and a cloth fiber is manufactured from the vegetable.

ONIONS

The onion has been known by man for over four thousand years, which accounts for doubt about its place of origin, probably Asia Minor. There is no doubt, though, about its popularity and even reverence at certain times in history. It was worshiped in Egypt long before the time of Christ and considered such a necessity for the workmen building the great pyramids that nine tons of gold, reportedly, was paid for the sweet onions (and garlic) they ate on the job. Egyptian tomb paintings have representations of onions, and an actual onion has been found in a mummy's hand. Egyptians could swear oaths on it, and the spherical nature of the onion made it an obvious symbol: "In the onion the Egyptians symbolized the universe, since in their cosmogony the various spheres of hell, earth, and heaven were concentric, like its layers" (Charles M. Skinner). The onion is sacred to Saint Thomas and was used for divining purposes by brides-to-be in England on Saint Thomas' Eve (December 21). It has been carried on the person and hung in the house for good luck and to avert disease. (A substance called allyl aldehyde in onions is antiseptic and will attract and destroy certain bacteria.) For epidemics a cut onion was hung in a room and the infected part was burned periodically. The odors of a newly painted room can be eliminated in the same way. In Yiddish "to grow like an onion,"

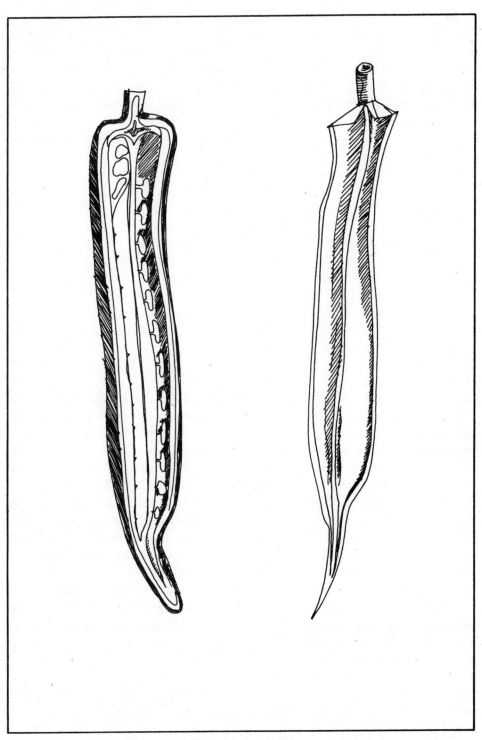
Whole fresh okra pods are sometimes found, but it is usually sold sliced, canned or frozen.

is an insult but it is certainly a compliment to "know one's onions."

"The Onion in its satin wrappings is among the most beautiful of vegetables, and it is the one which represents the essence of things; it can almost be said to have a soul. You take off coat after coat, and the Onion is still there; and when the last is removed, who dare say that the Onion itself is destroyed? though you can cry over its departed spirit!"
Charles Dudley Warner

Peer Gynt, in another tone:

> "You're no Emperor. You're just an onion.
> Now then, little Peer, I'm going to peel you,
> And you won't escape by weeping or praying. . . .
> What a terrible lot of layers there are!
> Surely I'll soon get down to the heart?
> (*Pulls the whole onion to pieces.*)
> No—there isn't one! Just a series of shells
> All the way through, getting smaller and smaller!
> Nature is witty!"
Ibsen

Onions are "kitchen lilies" and belong to a botanical family that is a cross between the lily and the amaryllis. It stores food in its bulb during the first year and, if left another year, will bear white-green flowers. The most common type is the single bulb of the seed-bearing onion, which grows beneath the soil. (The name derives from the Latin *unio,* meaning "one, single.") Another type is the green or Welsh onion; it comes from China and Japan, where it is known as the Japanese leek, and not from Wales—the name is a corruption of the German word *welsche,* meaning "foreign." This onion is referred to, sometimes, as a scallion because it does not have a single bulb but several small scallions (scallions were originally called scullions, loosely meaning "good for nothing"). The spring onion, of which the white Lisbon is the most popular variety, is a close relative. Both varieties have long-lasting green stems, which are also eaten, usually chopped up in salads. Tree or Egyptian onions have very small bulbs, which grow at the top of the stalk instead of flowers and seeds, and because of their size have no commercial value. The Canada onion, which also has very small bulbs, has been cultivated especially for pickling and for the drinker's surprise: the cocktail onion. Many varieties of onions are grown—red, white, and yellow; round and long (a Japanese type grows to a foot long).

Possibly onions were introduced into England by Caesar's armies, although wild ones grew there before that time, and into the West Indies, and from there throughout the Americas, by the Spaniards. Spanish onions are very popular. They are sweet and mucilaginous, like the Egyptian ones, which sometimes can be munched raw like an apple. (Onions grown in warmer climates are generally sweeter and milder than ones grown in less temperate areas.) More recently the large Bermuda onion, similar to the Spanish, has been developed and is especially valued for its sweet taste. Onions are grown in all temperate regions of the world and there are very definite seasonal varieties; summer ones will not ripen when planted in winter, and vice versa. In the United States, Texas, California, Indiana, and New York are among the states cultivating the most onions.

> "Mine eyes smell onions;
> I shall weep anon."
>
> Shakespeare

> "Too often the poet sees but the tears that live in an onion, not the smiles."
>
> Anonymous

It is the sharp, volatile, sulfurous white oil that onions contain more than the other members of the family that causes the involuntary tearing of the eyes. The Greek name for onion had to do with shutting the eyes while eating, and everyone knows what an excellent means it is to create artificial crying (remember the fairy tale about the princess who couldn't cry?). This oil is also responsible for the distinctive and, to many, offensive odor and aftertaste of raw onions (the oil is dissipated during cooking). Surprisingly, when Luther Burbank developed an onion without the oil, many people wouldn't buy or use it. There are a myriad number of techniques to reduce eye-tearing while cutting onions, such as running cold water over the onion before slicing, or cutting the onion without removing the stem end, or keeping one's mouth shut tight and breathing only through the nose.

An old proverb says "Eat onions in May / And all year after physicians may play." The onion has a long history as a home remedy. As noted, it contains a strong antiseptic agent and has been used for burns (cut a raw onion in half, warm it slightly, and bind it to the affected part) and beestings; by the American Indians in a poultice for the treatment of cancer; rubbed on the head to cure baldness. Since it promotes perspiration, it has been made into gruels and administered to persons suffering from lung congestion, colds, and chills (an old-fashioned chest-cold remedy is a syrup made from onion juice and honey). A roasted onion is a common remedy, even today, for earache and neuralgia. Onions are soporific and may induce sleep. Because it increases internal heat, the onion was considered a good tonic for phlegmatic personalities; conversely, it is not advised in feverish conditions or for people who have high body temperatures. Possibly this quality is the one that caused the Brahmans in India to shun the onion and generally prohibit it in yogic diets, along with other heat-producing food, as they aim to help the aspirant to achieve inner quiet. Negative views on onions include those of some hygienists who claim that their antiseptic properties serve to kill necessary intestinal flora, and test results show that excessive consumption may cause anemia or induce melancholy. As onions tend to draw disease to them, it is advisable not to eat any that show signs of decay. Some miscellaneous uses for onions include eating them as a preventive for thirst, using them as a nonpoisonous garden spray by putting them through a blender and then straining the bulk out; making a natural yellow dye by boiling dried peels.

Wild onions grow abundantly throughout most of the world. In the United States they can be found in every state and there are no poisonous species. They are easily recognized by their characteristic odor. Rocambole is a type of onion found growing wild in dark places along the coasts of northern England and Scotland. Sometimes it is cultivated. It resembles a purple-skinned garlic and has a mild garlic taste. Its other names are sand leek and rock leek.

PARSLEY

> Can you make me a cambrick shirt
> Parsley, sage, rosemary and thyme . . .
> *Gammer Gurton's Garland*

Probably this is the earliest recording of this famous refrain. The reason these herbs are frequently grouped together is that they were thought to possess magical properties—a fact that would make them a natural refrain for poems and songs. Actually, parsley is an herb, but it can (or should) be used more extensively in the kitchen, and for this reason I have chosen to include it here. Parsley was greatly favored by the ancient Greeks. At their lavish feasts they would crown their brows with the leaves, believing that it would excite the brain, increasing gaiety and appetite. In Homer's *Odyssey,* Calypso's island was planted with parsley. Parsley wreaths crowned the champions of the Nemean games, which were held in honor of a young boy, Archemorus. He was killed by a serpent, and it was believed that parsley sprang from his blood. By the same association, parsley was one of the sacred herbs of burial. It was placed on graves and the Romans served it at funeral banquets. Probably it was this custom that precluded the use of parsley at the table. Instead, it was used as a medicine.

"First he [Peter Rabbit] ate some lettuce and some French beans and then he ate some radishes and then feeling rather sick, he went to look for some parsley."

Beatrix Potter

According to Nicholas Culpeper, it is one of the five "opening roots"—a remedy for obstructions in the stomach and elsewhere. Galen used boiled parsley roots for epilepsy and John Gerard recommended it for ridding poison from snakebite. It has served as an eyewash, and the crushed seeds were used to eliminate freckles and lice. It is probably one of the best natural breath fresheners and will remove all trace of garlic and onions. Parsley tea is a good tonic, especially for rheumatism, and it also helps to clear the complexion. Science in modern times has provided explanations for ancient herbal remedies: parsley contains far more vitamin A per ounce than carrots (as much as cod-liver oil), three times as much vitamin C as oranges, and twice as much iron as spinach. This should be reason enough to put parsley raw or slightly cooked in salads, soups, and many other dishes, and not relegate it to the position of a garnish that is generally cast aside, uneaten. Parsley is sought out by rabbits and sheep as a cure for foot rot, but, paradoxically, it is poisonous to birds and chickens. (Fool's parsley is a wild herb that can be poisonous, although wild parsley has the same properties as the garden variety and can be eaten.)

Cultivated varieties of parsley are usually curly or fern-leaved (Italian) and not the earlier straight, divided-leaf type. Probably of southern European origin, parsley was cultivated by the Greeks as a garden border together with rue. The old saying "We are at parsley and rue" probably derives from this custom and means being at the onset of a project. The Jews eat parsley at the Passover feast; it symbolizes the arrival of spring and rebirth. Similarly, it is a sign of fertility wound around a carrot. (It is a generally compatible plant and a good neighbor to everything from tomatoes to roses.) There is even some reference to its use as an aphrodisiac. Conversely, in the Dark Ages there was a superstition that to transplant parsley would be to invite crop failure or even death. And there is an expression to the effect that "parsley, before it is born, is seen of the devil nine times," but this may also refer to the fact that Italian parsley grows so slowly that it "goes to the devil and back nine times before it comes up."

PARSNIPS

Parsnips are of the same family as carrots. This root vegetable grew wild (and still does) in parts of Europe and the Caucasus long before its cultivation, probably in ancient Rome. The parsnip is a fleshy root that is highly nutritious and has such a high starch and sugar content (the starch converts to sugar if the root is exposed to frost) that some people actually complain that it is too sweet for a vegetable! Parsnips have carried some strange superstitions with them: keeping one on one's person was said to ward off snakebite, but if you happened to forget, and get bitten, you could, according to the Greeks, crush a parsnip and mix it with the pork fat they used to grease their chariot wheels, spread the paste on the wound, and be cured. Many people thought parsnips were dangerous—especially old ones, which they believed could cause insanity. Native water hemlock, which grows wild in the United States, resembles parsnip and *is* poisonous. Medicinally, parsnips were used for intestinal problems by the Romans, and it has been reported that the emperor Tiberius had them imported from Germany so that his cooks could boil and serve them in a honey sauce. Around the middle of the sixteenth century in Europe, the parsnip became a common vegetable for the poorer class, much as the potato was. The English introduced them into America, where the Indians also learned to grow them. The usual type of parsnip is long and dingy white and has stems and leaves resembling celery. There are round parsnips, but these are not grown in the United States. As noted, the vegetable's sweetness improves when exposed to cold. The usual way to serve parsnips is steamed (not boiled, please!) and buttered; but they make an interesting addition cold in salads and are also made into wine. As their flavor is very distinct and dominating, it is not advisable to cook them in stews and soups. Edible wild parsnips grow in the southwestern United States, where they are called *gamotes*. Caution should be observed when identifying and eating wild parsnips.

An interesting note: The name of parsnip in Russian is Pasternak.

PEAS

>Pease Porridge hot,
>Pease Porridge cold,
>Pease Porridge in the Pot
>   Nine Days Old

This Mother Goose rhyme, which children chanted while clapping their

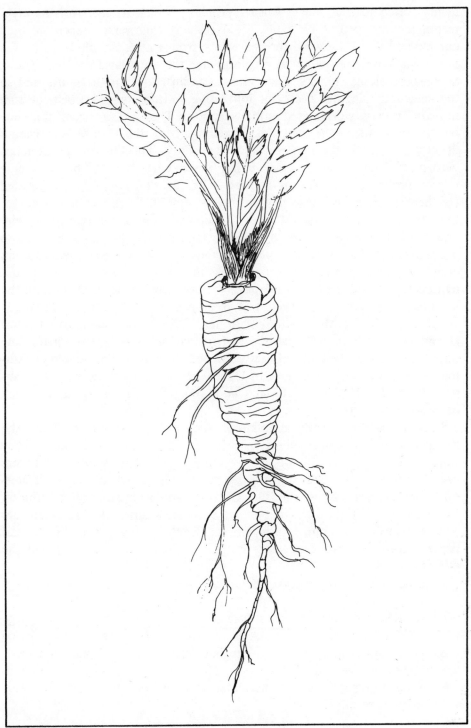
Parsnip looks like an off-white carrot and is very sweet.

hands (often to keep them warm), probably seems strange with its eighteenth-century spelling of "pease," but actually this is more accurate than our modern one. The word for pea comes from Latin; the plural form was *pisa,* which became *pesen* in English, and the singular form was *pease.* Somewhere along the way, the *s* sound was mistaken for the plural and so the inaccurate singular *pea* came into being. Garden or green peas (*Pisum sativum*) probably originated in the Near East. They are so ancient that they have been found in the Swiss lake dwellings from the Bronze Age (about 3000 B.C.) and in the remains of ancient Troy. The Greeks and Romans cultivated them (wild *P. sativum* has never been found), but all these people probably ate peas in a dried state, as they were first cultivated for their dry seed. Because they are seeds, equipped with nourishment for the next generation of plants, peas, like other legumes, are rich in protein, and in dried form can be stored easily for later use. Actually, the fresh green pea is picked before it is fully ripe, when the sugar content is at the maximum and well justifies the term *sweet peas* (as the pea gets older, the sugar content decreases and the starch increases). Peas were not eaten fresh until the seventeenth century. The sweetest of green peas were the tiny *piselli novelli* which Catherine de Médicis took with her to Paris when she married Henri II and which the French adopted and named *petits pois.* During the late sixteenth and seventeenth centuries these peas became the vegetable of fashion, with a price tag to match. In Queen Elizabeth I's time they were imported into England from Holland and were "fit dainties for ladies; they came so far and cost so dear."

Peas appear abundantly in folklore. They were a favorite of Thor, the thundered, and are still eaten on Thor's day, or Thursday, in Germany. They were often connected with wooing, possibly as a fertility symbol, and were used for divination: to dream of a dry pea was a portent of a coming marriage; and everyone knows the Hans Christian Andersen story of the princess so sensitive she could feel a pea under all those mattresses. The references were not always complimentary. An old English saying went "Love and pease pottage are two dangerous things, / One breaks the heart and the other the belly."

Peas fall into several general categories.

Garden Peas, of which there are subclassifications such as marrowfats (large and wrinkled; most used for canning and freezing); sugar, China peas, or snow peas, an essential in oriental cuisine (eaten with the pod, and seeds are practically nonexistent); and the aforementioned *petits pois* (which have more nutritional value when eaten raw and can be served in salads with delicious results).

Field Peas, which are used for feeding livestock and the seeds of which are dried for use as split peas. One type is the pigeon pea, which originated in Africa and was cultivated in Egypt before 2000 B.C. Field peas have purple flowers and gray, yellow, or spotted seeds, whereas garden peas have white flowers and yellowish white or bluish green seeds. Some dried peas can be sprouted like other legumes.

Asparagus Peas grow mostly in southern Europe and bear no resemblance to asparagus. They don't look like other peas, either. Instead, they have hairy pods with prominent longitudinal ribs and brown seeds. The entire pod is eaten when it is about one inch long.

Cowpeas originated in Asia or Africa and came to America with the first slaves. They thrive in warm climates and are so common in the southern United States that when southerners say "peas" they usually mean cowpeas; garden peas are called English peas. Cowpeas are botanically closer to beans and can be eaten fresh, together with the pods when immature; more often they are dried, made into meal, or ground into a coffee substitute. The most popular type of cowpea, particularly in the South, is the blackeye or black-eyed pea, which grows on a vine and can be easily recognized by its black "eye" on a white seed. A favorite southern dish is black-eyed peas cooked with a piece of pork.

Chick-Peas are discussed under that heading.

PEPPERS

Garden peppers are an entirely different species from white and black pepper, used as a spice. The name turned out to be the same in English because Columbus, in search of a short route to the East Indies to facilitate commerce of valued condiments such as pepper, came to the West Indies instead and found the natives there growing plants that he mistook for the spice. Actually they were varieties of the botanical genus *Capsicum* (the pepper plant's botanical name is *Piper nigrum*). He brought specimens back to Europe and they were soon accepted as a worthy vegetable—mainly for their value as seasoning. Peppers are reportedly more than two thousand years old and have been found in Peruvian ruins of that age. The ancient Aztecs also cultivated them. Interestingly, although they are an American vegetable, they had to be introduced to the United States from Europe, where they have been eaten since the mid-sixteenth century.

There are many varieties of garden pepper, which fall into the two main categories of sweet and red (or hot). Sweet peppers are preferred and are a more important vegetable in the American diet. The most common type is the bell pepper, large and bell-shaped, which is often eaten while still immature and green or else as fully ripe and red. This sweet red pepper is known as a *pimiento* in Spain. Although Americans associate the word *pimiento* or *pimento* with the sweet red stuffing in olives and cheese, or roasted pimentos in jars and cans, in Spanish the word also denotes the fiery hot pepper, or chili—so be careful! Hot peppers are sometimes called chilies (from the Mexican Nahuatt Indian) and are particularly popular in countries with warm climates such as Spain, Mexico, and India. Hot peppers are green or red; the latter is often dried and ground into chili powder, an essential ingredient in chili con carne. (The spelling seems to change at whim; *chilies* may be spelled *chiles* or *chillies*.) Tabasco sauce is made from pickling chilies. Another long, bright-red pepper is cultivated in Hungary, where it is called *paprika*. This can be either pungent or mild and is usually ground into a powder, which is a must in such dishes as goulash and paprika chicken. The mild form is used less for taste than for color. The condiment cayenne pepper is ground from dried pungent red peppers. Other types of peppers include the small, thin Italian green pepper and the small, round, and very hot cherry pepper, which can be red, yellow, or purplish. Should you have doubts about how "hot" a pepper can be: some types are so powerful that they cause irritation to the hands of the pickers. Sweet peppers are a very versatile vegetable, raw or cooked. Both sweet and red peppers contain a large quantity of vitamin C; raw sweet red ones contain the most.

POTATOES

> Pray for peace and grace and spiritual food,
> For wisdom and guidance, for all these are good,
> But don't forget the potatoes.
>
> John Tyler Pette

> "Let the sky rain potatoes; let it thunder to the tune of Greensleeves."
>
> Shakespeare

There is no more important or universal vegetable than the common potato. This unassuming and often unattractive tuber has been at times so linked with man's life that if an item of food can be said to have influenced history, the potato can. The ancestors of the potato originated in the Peruvian Andes and were cultivated for many centuries before the Spaniards arrived. They were grown primarily to break up the soil so that growing maize would be easier. When the Spaniards discovered them there, in the early part of the sixteenth century, they were being dehydrated in the sun and pounded into flour, as needed. One of Pizarro's priests is credited with bringing the potato to Spain, even though Columbus is sometimes mentioned (it was the sweet potato and not the white one which Columbus found in the West Indies). For a reason that eludes me, the potato was known as a revitalizer for impotence, and sold for as much as $1,000 a pound! Even so, most Europeans feared them to be poisonous at first and were very slow to accept them as a food item. Some basis for this exists in the fact that potatoes belong to the same botanical family as deadly nightshade and under certain conditions, which will be described later, can be dangerous. It is possible that some of the early specimens were semipoisonous or at least very bitter. Many people also shunned potatoes because they were not mentioned in the Bible (particularly, the Scots, as late as 1728). Finally the Irish found a way around this by planting them on Good Friday and sprinkling the soil liberally with holy water. Despite all prejudices, the potato was in common use by the beginning of the nineteenth century, at which time it was also brought to the United States (there is an erroneous story that potatoes were found by some of Sir Walter Raleigh's party in what was then called the Virginias, but that was probably another kind of tuber (see *Jerusalem Artichokes*). The vegetable became a life-saving staple during the famines in France in the late eighteenth and early nineteenth centuries. Parmentier, a philanthropist, agriculturist, and great friend of the court at that time, was concerned with encouraging the peasants in France to use potatoes (possibly as a substitute for wheat—remember "Let them eat cake. . . ."). To that end, Marie Antoinette wore potato blossoms in her hair and King Louis XVI sported one on his lapel; meanwhile the fields outside of Paris were planted with potatoes and then heavily guarded, giving the impression that something valuable was growing there. The ruse worked and the peasants raided the fields. They adopted potatoes into their diet so completely that during the Revolution, with an ironic twist, they even turned the Tuileries into a potato field.

"Be eating one potato, peeling a second, have a third in your fist, and your eye on a fourth."

Irish proverb

Who can think of Ireland without potatoes? The most common white potato is also known as Irish. It seems inconceivable that the Irish originally rejected this vegetable, which became such a necessity in their diet that during the crop failures of the 1840s a great famine ensued, causing the mass emigration of the Irish to the United States. Probably the reason for its great popularity in that country is that potatoes are very easy to grow and will flourish even in the poorest soil. They are also extremely versatile. They can be baked, boiled, fried, mashed, and so on, and to the Irish they comprised the "almost perfect" meal when eaten with herring, supplying essential mineral salts, vitamins A, B, C, and D, carbohydrates, protein, and fats. This leads to the controversy regarding the nutritional value of the potato. Some people claim a person could live entirely on potatoes for an indefinite length of time, others that they should not be eaten under any circumstances. The negative view relies in part on the botanical nature of the vegetable. As previously mentioned, the potato is a member of the deadly nightshade family, which includes several poisonous species. The potato begins as a small knob on the underground root of the plant from which grow many weak branches with leaves, flowers, and then berries. After these berries ripen and fall, the tuber swells beneath the ground and becomes the edible potato. Whatever danger lies in potatoes, it is not in the tubers themselves (although those that have sprouted or are still green can be inedible) but in eating those parts above ground—flower, leaf, stem, and berry, which contain the poisonous substance solanine. (Queen Elizabeth I's chef threw out the tubers and cooked the leaves!) The danger also seems to be increased by exposure to light. Whatever poison might have been present in earlier species has been eliminated by agricultural improvements, such as planting special seed potatoes (instead of old tubers), which when planted grow shoots from their "eyes." The blight responsible for the famines served a purpose, because it probably helped wipe out the former undesirable type of potatoes. Many of the potato's adversaries do not choose the factor of possible toxins for their arguments, but rather say that the vegetable is just a bulk food, composed mainly of water (in some varieties as much as 80 percent) and carbohydrates, and of less value than other bulk foods (like wheat) containing far more protein. Purporters of a macrobiotic diet prohibit potatoes (as they do some other members of the nightshade family, namely eggplants and tomatoes), claiming that the Peruvians and other early peoples had to roast, boil, and pulverize the potato in order to make it edible. The dissenters are a minority voice; most people consider potatoes an important source of carbohydrates (of course, composition will differ depending on soil, or age; the percentage of starch increases as potatoes grow older and new potatoes contain more sugar and water). They are

The potato plant's roots swell at their tips to form the edible tubers.

held to have the dual function of being useful in reducing diets (provided no butter is used on them) and for weight-gain, as they are highly digestible and absorb fats well. There appears to be evidence of some vitamin C content, especially in the skins, and it is recommended that one should not peel potatoes, since many of the minerals and other nutrients are contained in the skin. Boiling and peeling may result in as much as 47 percent loss of vitamin C. Chilling also protects the vitamin C, and potatoes should be stored in a cool place. Reportedly, waxy potatoes contain the most protein and floury ones the most starch. Potatoes are considered one of the most alkaline vegetables; some nutritionists suggest eating them raw, if possible. The vegetable has other uses; because it has high absorption properties, it is useful in absorbing odors or salt from overly salted food, and it can also be used as a stain remover to clean pots, wood, silver—and even skin. An old method for removing wrinkles was to grate a potato and press a bit on the skin around the eyes for about five minutes; for rheumatism, one was to carry a potato in a pocket or hang a piece around the neck. One can also create a good flower holder for flower arranging or a potato man for a child to play with. Some potatoes are cultivated to produce starch, gum, alcohol (particularly the homemade Irish whiskey known as poteen, which is very potent, and schnapps), and glucose, and an all-black South American type is used to make a dye. Potatoes come in a wide range of shapes, sizes, and colors. The most common is the white (Irish) potato, with brown skin and white flesh. Red- (actually pink-) skinned potatoes are popular. Unfortunately, in the United States some of these may be treated with wax or toxic weed killers to make them redder. A Congo potato has purple pulp. Certain wild potatolike tubers grow throughout the United States. Arrowhead, wapatoo, or duck potatoes, measuring no more than one inch in diameter, can be found growing in the mud of shallow ponds and were a favorite of the American Indians, as were groundnuts, which the early Pilgrims depended upon. These grow on vines and most probably were the ones found in the Virginias. Groundnuts are very small, smooth, sweet, and turniplike. In the United States, the regions most famed for commercial potato cultivation are Idaho, Maine, and Long Island, New York.

    The word *potato* in English is probably a cross between the Haitian word *batata,* more accurately used for sweet potatoes, and the Peruvian word *papas,* which became *patata* in Spanish. The French call it a *pomme de terre,* and the literal translation "earth apple" was often used in eighteenth-century England. Colloquially it is called spud or murphy, and its use as everyday fare has resulted in such frequently used expressions as "small potatoes" and to "drop like a hot potato."

PUMPKINS

The vegetable of autumn, pumpkin is a member of the large family of gourds and squash and technically belongs to the summer squash species. The differentiation at times is very difficult; the largest "pumpkins" grown are usually squash, as is the better commercially canned pumpkin. An American vegetable, found growing in South and Central America and Mexico long before the Europeans "discovered" these places, pumpkins grow on vines, between ten and twenty feet long, and range in size from several inches to two feet in diameter. The name *pumpkin* comes from the Greek *pepon* (meaning "cooked in the sun") by way of the French *poumpon,* which became English *poumpion,* and then was given the diminutive suffix *-kin*—very surprising, considering how large pumpkins grow. In Elizabethan English, a fat loutish man was called a "bumpkin"; the Greeks named a person considered weak and soft in the head a *pumpion.* The pumpkin has been a longtime favorite of children. Their literature is filled with representations of the large round yellow vegetable: turned into a coach for Cinderella, eaten by Peter Peter (who then put his wife in the empty shell). Jack Pumpkinhead lives in the Land of Oz and in the comic strip *Peanuts,* Linus reveres "the Great Pumpkin." But more than anything, how could a child celebrate Halloween without a jack-o'-lantern grinning in the window to frighten all evil creatures away on that awesome night? (At least one pumpkin has rebelled at this characterization. A well-known children's song goes, "Oh, the glory of the Jack is in the candle, from the gatepost where it grins set up so high; and the glory of the turkey is the drumstick, but the glory of the pumpkin is the pie!") Indeed, the pumpkin performs so beautifully as pie-filling (particularly female ones, I'm told) that one often considers it more a fruit, although it is an excellent main-course vegetable—delicious, a fine source of sweetness for sugar-free diets, and good for the intestines.

Gourds and melons appear often in mythology. The pumpkin, specifically, was a symbol of fruitfulness (rebirth) and health in China, where it is still called the emperor of the garden. An unusual version of the deluge story tells about a wise man in India whose only son took sick and died. In order to remove the body, the man enclosed it in a huge pumpkin and carried it to the foot of a mountain. Some time later, he was in the region again and opened the pumpkin. To his amazement, a large number of fish and even a few whales flowed out, followed by enough water for them to swim in. When the people of the town heard of this, they hurried to catch the fish and accidentally broke the pumpkin into pieces, from each of which flowed a river. Soon the water became an ocean and the earth was covered.

The Mayan Indians used pumpkin sap for burns, but medicinally the seeds have primary importance. American Indians chewed them fresh or dried for kidney infections or intestinal parasites. By the middle of the nineteenth century, they were an official drug in the pharmacopoeia as a diuretic and worm remedy. Chewing pumpkin seeds has been a longtime folk remedy for prostate trouble; investigation has shown that in cultures where this custom has been practiced for generations, the potency and health of the gland has been preserved. The seeds are very rich in phosphorus, iron, and many B vitamins, including niacin, and are about 30 percent protein and 40 percent unsaturated fat. Pumpkin seeds (sometimes called pepitas) can be purchased raw or roasted, or you can create your own supply by saving them and either allowing them to dry naturally or drying them in the oven. The outer husk should be peeled off and just the kernel eaten, either plain or toasted in a pan with a bit of oil. They are a wonderful snack or complement to cooked dishes and salads.

RADISHES

Wild forms of radishes found growing in China attest to the oriental origin of this pungent vegetable. The radish was eaten in Egypt even before work began on the pyramids. The ancient Egyptians not only ate the root as a vegetable; they made oil from the seeds, and so revered the radish that pictures of it were carved into the walls of a temple at Karnak on the Nile. No doubt the builders of the pyramids munched radishes along with their favorite garlic and onions. The esteem in which radishes were held by the Greeks is exemplified by the custom of serving them in gold dishes, whereas beets and turnips warranted only silver and lead. A statue of a solid gold radish stood in a temple at Delphi. The Romans considered it a root par excellence. Later on in Europe, radishes became less an item in the kitchen than one with healing and even magical properties: a radish could help one detect the whereabouts of witches; it could help to cure madness and possession by demons; shingles, headaches, eye aches, pains in the joints, warts; it could quicken the wits, take away black-and-blue marks, and even prevent poisoning (eaten beforehand), although horseradish is more acclaimed for this latter use. Radishes have a high alkaline content and are said to be beneficial in kidney and bladder conditions. They also contain a fair amount of phosphorus and sulfur and are good for the liver. Radishes can be juiced. (*Recipe:* Blend together three or four radishes, one half onion, and pine-

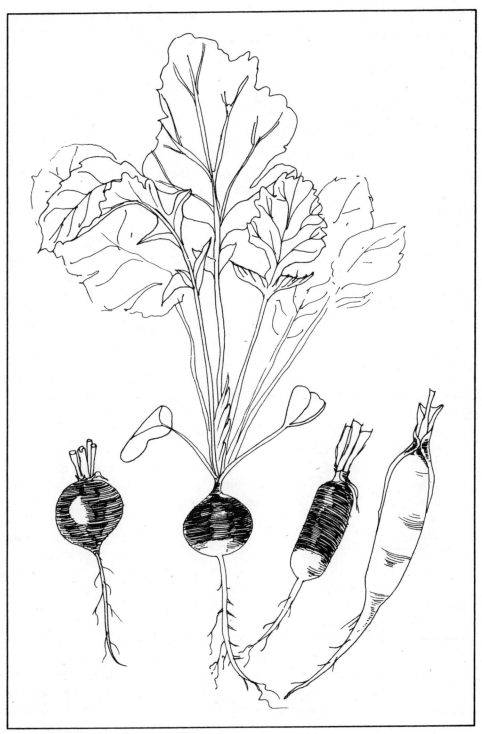
Radishes come in a variety of shapes and colors—red, white, or striped.

apple juice.) An old wives' method for making hair grow was to apply a combination radish juice and honey; radish juice could also be used to help remove corns.

A Talmudic story goes as follows: Once Judea was renowned for producing such enormous garden plants that a fox was able to hollow out a radish and make it his home. After he left it, curious onlookers weighed the vegetable and found it to be nearly a hundred pounds! This story is probably fictional, but it is a fact that a German botanist in the mid-sixteenth century reported seeing some radishes that weighed a hundred pounds. The radishes of antiquity grown in the Orient were much larger than our modern small salad varieties; in the East some cultivated radishes grow to as long as three feet. So-called winter (because they bloom late and can be stored like turnips through the winter) types like the black Spanish, China rose, and white Strasburg are fairly large. (For a discussion of different types of oriental radishes, see *Miscellaneous Oriental Vegetables*.) Common garden radishes are red or red and white (two attractive varieties of the latter are the cylindrical French breakfast and the round Sparkler). White radishes are less common, and black, like the black Spanish, are delicious, especially when cooked. Although most radishes are generally eaten raw, cooking removes much of the pungency some people consider to be unpleasant (radishes belong to the mustard family). Some are grown just for their leaves; in India there is a cabbage radish whose leaves are cooked like other greens. The root of the rat-tailed radish grown in southern Asia is not eaten; instead, the fruit, which grows to eight to ten inches long, is used raw or pickled in vinegar.

The English word *radish* comes from the Latin *radix,* meaning "root." The botanical name for the genus is *Raphanus* which is the Latin form of the Greek expression *raphanos,* meaning "easily raised," and accurate because radishes are an excellent home garden vegetable.

Horseradish, a root best known as a flavoring ingredient in sauces, originated in eastern Europe. The parts that grow aboveground are poisonous, as is the root when eaten in large doses (which is rare—it being a bit too strong to take). The French name for horseradish is *raifort* ("strong root"), and often it was known as hoarse radish. It was used widely as an antidote for various poisons, and one was to wash one's hands in the juice to remove any trace of snake venom. *Potter's New Cyclopaedia of Medicinal Herbs and Preparations* lists horseradish as a stimulant (particularly for the digestive organs), diaphoretic, and diuretic. It is also credited with being a remedy for rheumatism. Like radish, horseradish is a member of the mustard family, and it is one of the bitter herbs eaten at the Jewish Feast of

Passover. The root's pungency makes it an excellent accompaniment to many dishes, particularly those of north European origin. If you are grating it from the raw root, it is best to use a blender and be careful not to sniff; it will surely cause the eyes to tear.

RHUBARB

Rhubarb is not a fruit, although its common usage as a pie filling would seem to classify it as one; it is even called pie plant in some parts of the United States. Actually it is a member of the same family as sorrels, known as docks in England (one dock, the leaves of which were used for wrapping butter, is called rhubarb dock and is grown for its thick, succulent stalks). Possibly rhubarb derives from a wild Siberian species of the plant and is native to the cold regions of central Asia. The most popular type comes from a species that originated in the eastern Mediterranean region and Asia Minor. Garden rhubarb is a relative latecomer to the West; it was introduced into Italy in the early seventeenth century and brought to the United States around 1800. By the mid-nineteenth century it was popular, particularly in the New England states, as a pie and pastry filling (combined with sugar to counteract some of its natural tartness) or pressed into homemade wine. Rhubarb is a vegetable that grows best in northern climates, because it needs a long winter's rest. Under ideal conditions, the stalk will sometimes grow as high as three feet, with a width equal to a child's wrist; the leaves can grow as much as two and a half feet across. Stalks are round on one side, flat on the other, green or red; the latter is a later development and generally preferred. "Forced" rhubarb, grown early in the season, is more tender and less acid.

Nutritionally, rhubarb is one of the most controversial vegetables. Everyone agrees that its leaves should never be eaten, since they are very high in oxalic acid and have even been known to cause death. But some nutritionists go even further and say that the stalk also contains too much oxalic acid (which combines with calcium to form deposits in the kidneys) to be eaten and that there are not enough beneficial nutritional elements to justify its consumption, particularly by people who have a tendency to diseases of a genitourinary nature. Conversely, there are those who claim that rhubarb is excellent for the blood, the liver, and digestive disorders when eaten raw or juiced. It is an acid vegetable. Chinese rhubarb, which is entirely different from our garden variety, has been used medicinally since about 2700 B.C. The root is used as an astringent, tonic, and stomachic in small doses of powder. Large doses act as a purgative. Sometimes small pieces of the dried root are chewed to sweeten the breath and strengthen the stomach.

A curious use of the word *rhubarb* is when it is synonymous with a row or mob noises. This probably derives from its theatrical use; a piece of stage business supposedly created by Shakespeare and carried into the twentieth century was the custom of rhubarbing. In order to sound like a mob outside, a few actors would stand backstage and intone the word *rhubarb*. They were known as rhubarbers and the effect was astonishingly authentic. No one knows how it came to be, but if you try it you will see that it works.

RUTABAGAS

Rutabagas are an entirely different species from turnips, although the two are often grouped together. Actually, the rutabaga's hereditary structure indicates that it is a rare hybrid—a cross between a type of cabbage and a turnip. Whereas the turnip is an ancient vegetable, the rutabaga is relatively recent. There is no record of it until the middle of the seventeenth century. Rutabagas (*rotobagge* in Swedish) are often called Swedish turnips and in Europe are commonly known as swedes. They differ from turnips in appearance, as they have smooth cabbagelike leaves and a swollen neck with ridges, and although they can be purple, white, or yellow, the flesh is usually deep yellow or orange like that of a sweet potato, and used in the same manner. The growing season is longer than the turnip's, but the rutabaga's sensitivity to warmth is the same, so it grows best in northern regions. Rutabagas were used for food and fodder in France and southern Europe during the seventeenth century. The vegetable was introduced into the United States around 1800 and, although more nutritious than the turnip, it has never been very popular here.

SEAWEEDS

For those concerned with the question of a decreasing world food supply, seaweeds may provide one of the answers. These sea vegetables, as they are often called in the Orient, grow freely along rocky coasts throughout the world. They are high in calcium, sulfur, potassium, and especially iodine. The last-named is a very important element when there is danger of too much environmental radioactivity. (It has been reported that the Japanese who consume a great deal of seaweed have no thyroid cancer.) Once eaten more extensively in the West, especially in England, seaweeds have become less common; but already noted environmental conditions have brought about recent interest in phycology (the study of seaweeds) and greater seaweed consumption. Kelp, particularly in powdered form, is used like salt or taken in pills. Most common sources of seaweed are extracts. The Chinese and Japanese have always eaten them; in fact, seaweeds constitute about 25 percent of their diets and there are over a hundred species growing in the Pacific. The first written record of seaweed was in 600 B.C. in China. The ancient Polynesians planted "sea gardens" to domesticate their supply, and seaweed cultivation has been an industry in Japan since the early eighteenth century. Twig bundles are planted in clear saltwater areas and seaweed attaches itself to the twigs; it is then harvested and washed in fresh water. In the Orient, seaweed is cooked in soups and stews, wrapped around cooked rice balls, pickled, preserved, or dried. One type of red algae seaweed, laver, is common fare in Ireland and South Wales, where it surprisingly appears as a breakfast food—coated with oatmeal and fried. This thin, wavy-edged, reddish purple seaweed becomes olive green or brown when cooked and is often sold in these countries (and parts of Europe) boiled and pureed like brown spinach, or is seen cooking in the "sloak kettles" of Ireland and Scotland.

Seaweed is more often sold dried than fresh. Most types require little cooking (tougher ones have to be soaked and cooked longer). They may be added to salads, main courses, soups, and rice, eaten as snacks, or crumbled or powdered as seasoning. Some of the types best known in the West are:

Murlins, which are brown seaweeds. The most popular murlin is known as tangle. It is eaten raw, mainly in Ireland and Scotland.

Dulse is a popular dark-red member of the red algae family, which is sometimes found fresh (especially in Canada, where the children eat it like candy). Tougher dulse is usually washed and then dried for cooking or chewing (like gum). In New England it is sometimes served as a relish. Like most seaweeds, dulse is very salty and requires no additional seasoning.

Irish Moss (Carrageen) is used commercially for the gelatinous substance, carrageenin, that is extracted from it. This is added to jellies, ice cream, soups, and cosmetics as a natural emulsifier. Also red in its fresh state, Irish moss is used powdered in the famous dish blancmange. Medicinally, as a tea, it is a remedy for coughs, ulcers, diarrhea, and kidney and bladder disorders.

Kelp, as previously mentioned, is the general name for large brown algae. Brown seaweeds are less palatable than red and are usually made into powders or mixed with meal for feeding to livestock.

Agar-Agar is similar to Irish moss. It is found often in translucent strips or fine white powder and is an ideal natural gelatin that can be substituted for commercial gelatins. Additionally, it is an aid in organic disorders such as those of the liver; or tuberculosis; or gout. Medicinally, it is used best dry and mixed with juice or another liquid.

Samphire grows on seacliffs along the coasts of the British Isles, the Black and Mediterranean seas, and the Atlantic Ocean. It is an aromatic plant with thick fleshy leaves and can be eaten baked or steamed quickly, pickled, or raw in salads. It is supposed to be beneficial to the kidneys and skin.

SHALLOTS

Shallots were once thought to be a separate species but now are acknowledged as a member of the onion family. These small bulbs look and taste like a cross between onions and garlic. Their skins are orange-brown; their structure is garliclike cloves. The origin of shallots is not known. The crusading knights brought them back to their homes in England and France from Syria during the twelfth century, and they acquired the name of *eschalot* in French from an erroneous idea that they came from Askalon in Palestine. They are still known by that name, as well as *cibols*. Shallots have always been a favorite with the French, the delicate flavor providing just the right "touch" to soups, pâtés, and so on. Sometimes shallot cloves are rubbed onto bowls in the same manner as garlic, or used dried. The French brought them to Louisiana, where they often appear in the local cuisine. Shallots are a good home garden vegetable and there is a proverb to assist the planter: "Plant on the shortest day, lift on the longest day."

SPINACH

Although there is a record of spinach being introduced into China as early as A.D. 647, Europeans did not know the vegetable until the eleventh century. Spinach originated in Persia, where it was cultivated beginning in the

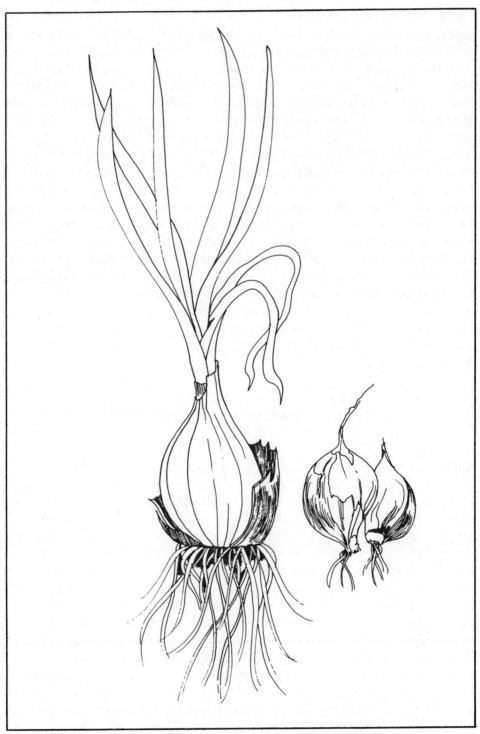
A member of the onion family, the shallot has individual cloves like garlic.

fourth century. Not mentioned in the writings of the early Greeks and Romans, spinach took another route, via the Moors, who acquired such a taste for it that they named it the "prince of vegetables" and brought it to Spain along with their many other influences. (The Crusaders returning home to England from Asia Minor also brought spinach back with them.) The etymology of the word for spinach is indicative of its journey: in Persian it was *ispanai,* which became *isbanakh* in Arabic, *espinaca* in Spanish, and *espinage* in medieval French. In many parts of Europe it was known as the Spanish vegetable and during the Middle Ages was one of the foods that the monks ate on fast days. The original spinach plant was spiny but became leafy as it adapted to growing in colder climates. There are two main types of spinach: round-seeded (summer) and prickly seeded (winter). The latter was an earlier development; the former was cultivated only since the mid-sixteenth century and is now generally preferred. Although the leaves are the part of the vegetable usually eaten, many species are valued for their seeds. Spinach is a member of the goosefoot family and has many close relatives, most of them so-called weeds, such as pigweed and amaranthus spinach, that can be used as substitutes. Other varieties include the spinach beet grown for its spinachlike leaves (its root is unimportant), and the popular New Zealand spinach, which belongs to the ice plant family but is eaten like spinach.

> "I say it's spinach and I say the hell with it."
> Cartoon by Carl Rose in *The New Yorker* magazine

Spinach came to the United States during colonial times and became a popular vegetable by the nineteenth century, but it took the comic-strip character Popeye the Sailor to make it an almost mandatory food for children. They were taught through Popeye's words and muscles that spinach was *the* food to make them strong, and at least one generation of mothers shoveled forkfuls into reluctant mouths to the tune of "Popeye the Sailorman." But some nutritionists choose to be contrary. Acknowledging that spinach is rich in iron, copper, vitamins A and D, and calcium, they argue that the vegetable contains too much oxalic acid. Oxalic acid has been discussed under *Rhubarb;* the problem arising from its presence in spinach is that when it combines with the vegetable's calcium, it tends to render that element useless. Even those most adamant on this point do not condemn spinach as being harmful; rather, they suggest that it be eaten in moderation, and not as a main source of calcium. The best way to eat it is raw (it makes a delicious salad green) or steamed quickly in its own juice, the water left on the leaves after washing—briefly. Soaking spinach may eliminate all the sand, but will undoubtedly also eliminate all the vitamins. There is a great deal of nutritional value to compensate for the oxalic acid, and spinach is an excellent blood-builder. Even more important, spinach *can* taste good. When well prepared, not cooked too much and overmixed with adulterating sauces, it can overcome any residual stigma from our youth. There are many gourmet dishes featuring spinach; the word *Florentine* in a recipe usually denotes its presence.

SQUASH

Squash belong to a large botanical family comprising gourds, melons, cucumbers, and pumpkins. They are distinctly American, originating in Central America and Mexico, possibly five thousand years ago. The American Indians grew them throughout this country and they were not even known in Europe until about the beginning of the sixteenth century. The word *squash* comes from the Algonquin Indian word *askootasquash,* which means "eaten green or raw," although today squash is usually eaten cooked. The entire genus name is *Cucurbita;* all are trailing and climbing plants with yellow, unisexual flowers. Edible varieties fall into two main categories: summer and winter. All summer squash are varieties of the species called *Cucurbita pepo;* they are quick-growing and should be picked and eaten while still immature, before the rinds or seeds harden. Outside the United States they are generally referred to as vegetable marrows. They are usually larger and more symmetrical in shape than winter squash. Some of the more popular types are summer crookneck (as descriptive a name as can be imagined) and summer straightneck, both of which are bright yellow or orange; Italian marrows, more commonly known as zucchini (*courgettes* in some places on the Continent), which were developed to be used whole when very immature and only a few inches long; longer ones may be sliced or stuffed; and custard marrows or scalloped summer squash (sometimes called cymling in the southern United States), which resemble a scalloped white or whitish-green cushion. Pumpkins technically belong to the summer category, but usually all late-growing, large, smooth, and symmetrical squash are called pumpkins, regardless of their species. Winter squash is usually picked when mature in the autumn and can be stored for three or four months into winter. They are firmer and less regular in shape and usually larger (except for the small acorn, which more resembles summer squash). Some varieties have rough or warty skin. Winter squash is supposed to contain more vitamin A, protein, fat, and carbohydrate and less water. Some better-known varieties are acorn, cushaw (often striped), and the large Hubbard, which is so floury that it makes delicious pancakes when grated, like a potato, mixed with egg, and deep fried. Squash blossoms are considered a delicious delicacy and can be stuffed or dipped in batter and deep-fried, the Italian way. Connoisseurs say male blossoms are best for this, but you'll have to discover for yourself how to tell the sex. A type of small courgette grown in the Middle East has a large flower and a small bulb (vegetable) at the end. It is cooked in its entirety and can be eaten in one forkful. Squash seeds, like those of the pumpkin, are highly nutritious. Other kinds of lesser-known squash are the chayote, which is grown in the West Indies (the whole plant—fruit, shoots, leaves, and roots—is used; the root can weigh as much as twenty pounds and looks and tastes like a yam), and the white or wax gourd, which is eaten extensively in the East.

There are many kinds of gourds, some inedible and used as decorations since ancient times or those such as the bottle gourd grown mainly for their hard shells, which are used as containers for cooking or serving. The snake gourd is a long, snakelike vegetable, sometimes edible; some others are grown for their seeds from which oil is pressed; some are particularly fibrous when mature, and one, the loofah, is raised to make into a natural scrub brush. The gourd has been a symbol in many religions. It was the primal egg of the Hindus; to the Hebrews Jonah's gourd ("which came up in a night and perished in a night") taught him compassion. Gourds were used to point the moral of rapid growth and quick decay, and they illustrate many religious emblems. In Christian symbolism, a gourd is the attribute of Christ, Saint James, and the Archangel Raphael.

SWEET POTATOES

Except for the similar size of their tubers, sweet potatoes and white potatoes are not related. The sweet potato has more claim to the name because it was known as *batata* in San Domingo, where Columbus first found the plants. Sweet potatoes originated in South and Central America, where the Incas and Mayas grew them to produce dyes for paints, as well as for eating. About one thousand species of sweet potatoes exist. The vegetable thrives in a hot, moist environment and is a staple food in many tropical and subtropical regions. Cultivation began in Europe during the sixteenth century, and when the early colonists settled in Virginia in the seventeenth century, sweet potatoes were one of their crops. Today they are preferred in the southern United States.

This winter squash is called Turk's Turban.

Sweet potatoes are elongated or spherical tubers. The skin can be white, light red (in the United States, red ones are dyed sometimes, and when washed will actually "bleed"), purple, or many shades of brown. The flesh is white or yellow, the latter favored in the United States and higher in vitamin A. Sweet potatoes contain some protein, and their starch content is higher than that of white ones. Generally they contain less water and, unlike white potatoes, do not store well. In addition to their use as a vegetable, their sweetness (up to 10 percent sugar) flavors pies, puddings, ice cream, and even candy (in Japan). In the Orient a flour is made from ground dried tubers. Sweet potato leaves are not poisonous as the white are, and they are sometimes used as potherbs. Sweet potatoes generally fall into two main categories: those with hard, dry, yellow flesh grown in the north, and those with soft, sweet, moist yellow or white flesh grown in the south and often, erroneously, called yams (yams are a totally different species and rarely seen in the United States; the Louisiana yam was so named to distinguish it from other varieties of sweet potatoes). Paradoxically, the soft sweet potatoes contain less water than the dry ones.

Sweet potatoes make an attractive house plant. Some are sprayed with growth inhibitors to prolong shelf life and will not sprout; but by trial and error, you can find one that will. Partially immerse it in a glass jar filled with water and when the shoots appear, plant it in rich soil with a bit of sand. It grows fast and may even bloom with triumpet-shaped blossoms, resembling its relative, the morning glory.

TOMATOES

Invariably, there is the question of whether a tomato is a fruit or a vegetable. The botanical answer is that it is neither; it is a berry. But in the United States it has been officially designated a vegetable by the Supreme Court. The ruling came as a result of a case in 1893 involving an importer who claimed that tomatoes were a fruit and therefore not subject to duty. He lost; the Supreme Court decided that the tomato's place in soups and main courses entitled it to be called a vegetable.

Tomatoes were cultivated since pre-Columbian times by the predecessors of the various Indian races of South America, and wild ones still grow in the lower altitudes of the Andes. One of the earliest mentions of the seeds was in A.D. 200 by a Roman, but in those days it was largely an herbal curiosity. The Spanish conquerors brought back *tomatl* (Spanish: *tomate*) seeds, which were planted in Morocco. The vegetable traveled from North Africa to Italy, where it was named *pomo dei mori* ("apple of the Moors")—modern Italian, *pomodoro,* which was somehow confused into *pome d'ore* ("golden apple") by the French and then further adulterated into *pomme d'amour* because, along the way, it acquired a reputation as an aphrodisiac, probably because it wasn't eaten as food in those days. The name stuck, and the tomato came back to America as a "love apple." (The Germans had another idea; they called it *Paradiesapfel,* for they believed the Turks brought it from the Holy Land.) Tomatoes were wrapped in superstition. Like the forbidden potato, they also belong to the deadly nightshade family. But they were certainly more attractive with their bright shining colors; so, although they were not eaten, they were used as decorations in gardens and on tables. In India, they were treated as harmful objects by the Brahmans and the British adopted the prohibition, probably easily, from a puritanical viewpoint that how could something that looked so luscious be anything but sinful! (In line with that reasoning, it is no wonder that often the "love apple" became the "mad apple.") It seems quite incredible that it took the Italians, whose cuisine in some regions relies on tomatoes, two centuries to eat the vegetable; but when they began, they did much to improve the cherry-sized yellow tomato and bring it to its modern large red form. (The smaller Italian plum tomato is the base for their thick pastes and sauces.) Thomas Jefferson was one of the first Americans to grow tomatoes as food, in 1781, and they were used as a vegetable by the French in New Orleans in 1812. By the middle of the nineteenth century, acceptance was universal.

Tomatoes require a temperate climate, but by careful, selective breeding, varieties were developed that were less perishable. By the time tomato juice became popular (in the 1930s), people had discovered the tomato's versatility and it became the "newest" important vegetable. Tomatoes are used in every stage of growth: when green they are pickled; when ripe they are eaten raw, stewed, baked, broiled, boiled, and so on. Some people favor eating them with the skins and seeds removed. In some parts of the Mediterranean region, they are cut in half, dried in the sun, and preserved in olive oil; in French cuisine, the term *Duglère* denotes their inclusion in any recipe. Far from being poisonous, tomatoes are rich in vitamin C (particularly those vine-ripened naturally in the sun) and contain vitamin A, the amount differing with soil and weather. They are also high in potassium. As with potatoes, tomato leaves and stems should not be eaten because of their unpleasant flavor.

There are many kinds of tomatoes: egg-shaped (which sometimes grow to be as heavy as one pound); cherry, which are small, red or yellow, grow in clusters, and are very decorative for salads and as garnishes; pear-shaped; the most common round, red ones, which grow quite large, like the popular beefsteak; and rare tree tomatoes, which belong to the same family but grow on trees instead of vines.

Remember the days when a particularly luscious-looking girl was referred to as a "tomato"?

TRUFFLES (see Mushrooms)

TURNIPS

There is a story of an ancient king who was exceedingly fond of a certain fish and requested it of his chef. The man was unable to find it and instead found a large white turnip, carved it into the shape of the fish, baked it, and served it to the king on a fish platter. The king was delighted and exclaimed that it was the best fish he had ever eaten!

Turnips date from prehistory and were a staple food for many of the early races, principally the Gauls. The Greeks had several varieties and even made small lead replicas of the root vegetable. There's an ancient Roman recipe that reads as follows: Boil turnips and strain out the water; then add cumin, rue, and benzoin and pound well in a mortar. Afterward, add honey, vinegar, gravy, boiled grapes, and a bit of oil. Simmer.

The turnip is extremely hardy and grows easily in sandy soil where little else thrives. It grew in the area from the Mediterranean to Asia Minor. There are records of turnips growing in France during the first century A.D. and in Henry VII's England, where the roots were baked or boiled and the tops were eaten like other greens. The European varieties of turnip that developed in the Mediterranean region are the ones commonly grown in the United States. Turnips were first brought to Canada in the mid-sixteenth century and planted in Virginia and Massachusetts when these colonies were settled.

Turnips were cultivated from field cabbage; their leaves are cabbagelike but rough and a bit hairy. Turnips can be white, yellow, red, or gray; long, round, or flat. The common ones have white or yellow flesh, with a green or purple band across the top of the skin. They can be very sweet. Turnips are an ideal winter vegetable and grow best in cooler regions. The roots, usually baked or cooked—alone or in stews and soups—or pickled for kraut, have little nutritional value, but the leaves, used as greens, are high in vitamins and minerals. Some kinds are cultivated only for their greens; others, called rape or colerape, for their seeds, from which an oil is pressed or used for birdseed.

Common turnips are white or yellow, often with a purple band.

In Roman times, turnips were known as *rapa* or *napus;* in Middle English they were called *nepe* or *naep,* which when combined with the Anglo-Saxon word *turn* (which might have meant "made round") became *turn-naep* and then *turnip.* In England, it was sometimes called a wort, like other cabbage. (The Anglo-Saxon word *wyrt* was a name for any root.) Turnips fell from favor for a while and, like tomatoes or eggs, were thrown as a sign of disapproval. A girl gave a turnip to a suitor in the equivalent of the cold shoulder. During the Middle Ages they became popular again and have remained so. In the southern United States they are common fare, particularly favored when cooked with pork. That chronicler Uncle Remus wrote, "Hog dunner w'ich part un'im'll season de turnip salad."

WATERCRESS AND RELATED GREENS

The history of watercress reverses the usual order. This excellent salad green was a cultivated plant and *then* began growing wild, not subject to cultivation on a large scale again until the early nineteenth century. The reason for this phenomenon may be found in the way watercress grows. In addition to sprouting from seed, watercress will root easily if you just toss a few stems into running fresh water. If the spot is favorable, the plant will not only flourish, it will spread. One can buy cultivated watercress in a store, but it's hardly necessary since it grows wild in every state of the United States. The only concern need be that the water where it's growing is pure; if there is danger of pollution, it should not be used fresh but the leaves may be cooked and eaten. Watercress is so prolific that one plant may yield as many as ten crops a year and grow for ten years. It also has the longest season of any vegetable and can be gathered almost year round.

Watercress bears the botanical name of *Nasturtium,* but is not the same species as the flower; the name literally means "nose twister." The name is justifiable because the plant is a member of the mustard family and a bit pungent.

> "Eat cresses and get wit."
>
> Greek saying

Watercress was used widely as medicine; the Persians ate it with bread and liberally fed it to their children, as they believed it helped them grow; the ancient Greeks served it to their soldiers for their health; the Romans ate it to prevent falling hair; and it was included on several official lists of medicinal herbs. John Gerard recommended: "Water Cresse being boiled in wine or milke, and drunk for certain daies togither, is verie good against the scurvie or scorbute." Sometimes called scurvy grass, watercress is very high in vitamin C and calcium. It stimulates digestion, is a good blood builder, helps in endocrine deficiencies (as a treatment it can be put in a juicer with pineapple juice), and cleanses the skin. External applications of watercress juice can rid the skin of blemishes. It may be eaten raw in salads and dainty tea sandwiches, or cooked like spinach. It is the starring ingredient in a superb French soup called *potage cressonière*.

Cress is like mustard greens (see next paragraph). Pots containing soil already seeded can be purchased for home growing and it will take about three or four more days than mustard to sprout. A native of Asia, cress has crisp leaves like parsley and is also used as a garnish.

Mustard Greens are often combined with other cresses in salads and sandwiches. They are quite pungent (the seeds are the basis for the condiment mustard) but very nutritional, as they have a high vitamin and trace mineral content. Unlike spinach and most other greens, they require a long time in cooking. Mustard greens are usually grown in greenhouses but can be sprouted at home in little time (three to four days in the spring and autumn; six to seven days in the winter). Mustard greens belong to the large family of brassicas, like cabbage and cauliflower, and can be found growing wild. They should be gathered in early spring when they are at their mildest.

Winter Cress is a close relative of watercress. It is rarely cultivated but in the wild is so resistant to cold that it can be picked throughout the winter, to provide a fresh green addition to salads. Part of its botanical name is *Barbarea,* because it is the only green plant that can be picked on Saint Barbara's Day (December 4). Winter cress is known variously as spring cress, upland cress, or yellow rocket. When it is cultivated commercially, it is sold under the names of scurvy grass or Belle Isle cress. Winter cress is not bitter when picked while the weather is still cold. If the taste is too strong for raw use in salads, it may be blanched and the leaves and buds (which are like broccoli) may be cooked.

There are many other cresses. Some of their names are shepherd's purse, pepperwort (garden cress), and bitter cress. Few of them are commercially cultivated; most of them are pungent and some very hot.

YAMS

Yams are tropical roots, probably of African origin and rare to Europe and the United States (see *Sweet Potatoes*). There are over two hundred species, of which about ten are important as food. In some parts of the American, Asian, and West African tropics, yams are the staple food. They store better than other tropical roots, have a high starch content, and can be boiled, mashed, roasted, or fried. So important is the yam to the Fijians that their eleven-month calendar is based on the vegetable's growth cycle. Folklore and superstition center around the vegetable; on some of the islands, it is believed that the root can move from one's land to one's neighbor's and magic must be employed to retrieve it. In some places, natives exhibit their yam crops at markets with pride, even framing and painting decorations on the best ones. Inedible yams serve other purposes; one type is used to poison darts. The yam is so prevalent in these regions that many roots are called by its name, and the English word comes from the Guinean verb *nyami*—simply, "to eat." Yams have been cultivated in the Orient since ancient times, and there was an attempt to cultivate the Chinese yam in Ireland as a substitute for potatoes during the great famine, but the root burrowed too deep. Only one type of yam, known as cush-cush, is native to the Americas.

Yam bean is a name applied to several species of edible tubers, seeds, and pods.

MISCELLANEOUS ORIENTAL VEGETABLES

Most of the vegetables discussed below have an important place in the Japanese and Chinese cuisines, although they may be included in other kinds of dishes with delightful results. Some, such as bamboo shoots and water chestnuts, can be purchased canned; fresh greens are available in oriental markets. Greens should be cooked quickly, either in a bit of oil or steamed in their natural juices.

Bamboo Shoots are the thick pointed young shoots that grow beneath the bamboo plant to become new stems if not cut as a vegetable. The bamboo plant is symbolic of many attributes, including long life, strength, and grace, and serves a multitude of functions in the life of oriental people. Bamboo shoots are six inches to one foot tall and three inches in diameter. They are boiled, salted, or pickled. In the United States they are usually sold sliced and precooked, in cans.

Chinese Cabbages Several different varieties of greens are referred to, colloquially, as Chinese cabbage. To help clear the confusion, here is a brief description of the more common ones:
1) Pak-choi (bok choy): Some types of this green resemble the spinach beet; others, Swiss chard, with their elongated white stems (which grow from ten to twenty inches) and broad, smooth green leaves. The vegetable is unlike European cabbages in that it doesn't form a heart. It can be eaten raw or cooked.
2) Pe-tsai literally means "white vegetable." There are two types of Chinese cabbage known as pe-tsai. One has light-green veined leaves, broad midribs, and a head like cos lettuce. It is sometimes called (erroneously) Chinese mustard. The other looks more like romaine lettuce. It has a long, solid head of light-green or white crinkled leaves. While cooking, there is no cabbage odor. These vegetables taste a bit like celery and, because of the shape, are sometimes called celery cabbage. Pe-tsai was cultivated in China as early as the fifth century. Seeds were brought to Europe and then to the United States by missionaries. It can be planted in home gardens and thrives in cool weather. Pe-tsai may be eaten raw or cooked.

Daikon is a Japanese radish, the best known of the many varieties grown in the East. This long, white radish is more pungent than ordinary garden types and is often pickled or shredded for relish or thinly sliced in soups, salads, or sushi (raw fish dishes). Daikon is so popular in Japan that it accounts for 25 percent of that country's vegetable crop.

Gai-Choy is Chinese mustard. It is a pungent green that can be eaten raw or steamed. The root is eaten like celeriac; the seeds are used in the same manner as Indian mustard, but are not as hot.

Chinese Okra is much larger (up to a foot long) than other okra and tastes entirely different—crunchy and juicy. It is dark green, with deep ridges, and can be steamed or fried.

Gai-Lon is Chinese broccoli. Unlike common broccoli, it has more leaf than flower. It can be recognized by its long stem, at the end of which are large light-green leaves and small flower buds.

Lo Bok is the name for a type of Chinese turnip similar to daikon and used the same way in Chinese cuisine.

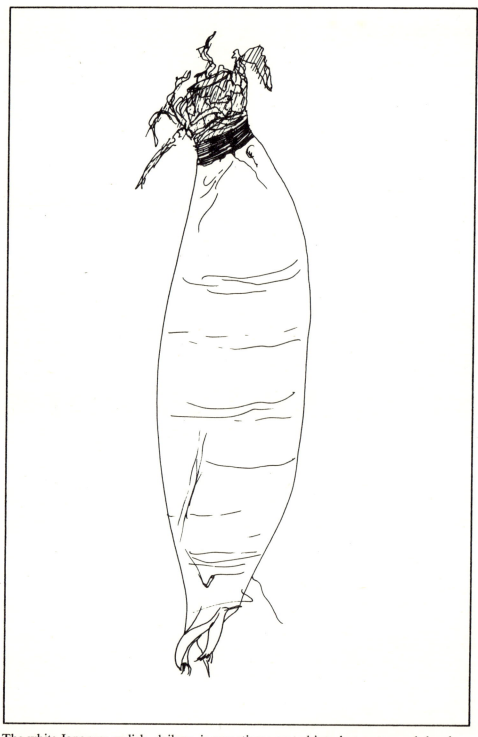
The white Japanese radish, daikon, is sometimes grated into hot water and drunk to reduce fever.

Shungiku is a type of chrysanthemum that is cooked as a vegetable in China and Japan. Its fleshy leaves have a very strong flavor.

Water Chestnuts are an ancient floating aquatic vegetable with a chestnut flavor. They may be eaten raw but are usually boiled, roasted, or made into flour. The Chinese water chestnut (pi-tsi) has a brown shell like a chestnut and white flesh that stays crisp and crunchy even after cooking. Its delicate flavor complements salads and cooked dishes. Generally, in the West, it is bought sliced and canned.

Some roots used in oriental cuisine are:

Burdock (Gobo) is a skinny root that grows as a weed here but has an important place in Japanese, particularly macrobiotic, cooking. The young stems can be eaten like asparagus, sautéed with other vegetables, or made into tea. Burdock has an earthy taste that takes some getting used to, but is supposed to be very therapeutic as a blood purifier and rheumatism remedy.

Ginger Root, although not confined to oriental cuisine, is used liberally for the pungent flavor it imparts to sauces and marinades. This knobby root is excellent mixed with fresh fruits and in baking; it is the basis for real ginger ale.

The Lotus is a type of water lily most sacred in the East. It is the symbol of life, resurrection, and immortality. In the plant all stages exist at the same time: seeds, buds, and flowers. The five-petaled flower grows in the mud to rise above it and face the sun. It is the symbol of meditation; the Buddha is often shown seated in the lotus position. So symbolic and holy is the lotus that one would doubt its use as a vegetable—but it is also eminently functional; every part is used. The leaves can be disposable plates, the seeds are eaten raw or roasted (like nuts), the shoots, as greens. Most important is the root. It is white, with spongelike holes in it. It may be eaten raw, boiled, roasted, or ground into flour. A tea made from the dried, powdered root is supposed to be efficacious as a remedy for respiratory ailments.

Japanese Seaweeds are numerous and very important in oriental diets. The best known are:
1) *Kombu* are brown, large seaweeds sometimes called oarweeds; they are processed and pressed into a dry cake known as *kombu*. *Japenese kelp* is often an ingredient in kombu. It is also powdered and used as a seasoning; pickled or candied; or added to sake (Japanese rice wine).

2) *Mirume* is small and dark green. It is eaten fresh, dried, or salted; usually prepared with vinegar and soy sauce.

3) *Hiziki* (hijiki) is black and stringy. If dried, when soaked it expands and is an excellent vegetable by itself—cooked with a bit of soy sauce, or added to other vegetables. It is sometimes called black rice.

4) *Wakame* is a type of murlin seaweed. It is dark, flat, and broad. It can be roasted or boiled (particularly in soups) or wrapped around rice cakes.

MISCELLANEOUS VEGETABLES

Hearts of Palm are the buds of certain palm trees. They are not cultivated in the United States but are imported, canned, as a delicacy for salads and side dishes. Sometimes known as palm cabbage, palm hearts are white and resemble artichoke hearts.

Nettles

> Nettles don't hurt
> If you count to ten. . . .
>
> A. A. Milne

The nettle is a plant with tiny hairs along its stems, which when touched can break the skin and burn it with the acid contained within. Its name means "spite"; and all together you would think it would best be avoided. But it shouldn't. If one is careful while picking to hold the plant in a way that the hairs face downward, it will be harmless. The tops and young shoots have no sting and are very nutritious. Nettles make a good potherb and can be cooked like spinach. (Actually, in eastern Europe, the sting was utilized; rheumatism sufferers had their backs flogged with nettles.)

"It is said that the Roman Nettle which thrives in England was originally planted there by Caesar's soldiers, who not having breeches thick enough to enable them to withstand the climate, suffered much in the cold, raw fogs; so when their legs were numb, they plucked nettles and gave those members such a scouring that they burned and smarted gloriously for a day."

Potter's New Cyclopaedia of Medicinal Herbs and Preparations

Potter's does not recommend this use, but does list the nettle as a diuretic

Florence fennel's leaves resemble the herb, but the licorice-tasting bulb is the part eaten.

and tonic. Other medicinal uses include stomach emetic, blood purifier (it is rich in iron and calcium), and poultice for shingles and wounds. After the top leaves are boiled for eating, the water can be used for soup or made into beer; and one type, patchouli, is a well-known aromatic oil.

Salsify has been eaten in southern Europe and North Africa for almost two thousand years. It has been cultivated since 1600 and was in the United States before the nineteenth century, but is not grown commercially to any extent. Salsify is sometimes called goatsbeard or oyster plant, the latter because of its taste, which some say resembles oysters. The plant has a flower like a huge dandelion, but it is the fleshy white root that is cooked and eaten like asparagus, or grated, coated with batter, and fried. It is a member of the chicory family and, when young, the leaves can be used in salads. Provided the ground doesn't freeze, salsify can remain underground and be picked throughout the winter as needed. Spanish and black salsify (*scorzonera*) must have their black rinds removed before eating. They are sweet and sometimes used as a substitute for coffee.

Sorrel is an herb, often used as a vegetable. It is quite acid and large amounts have been found to be harmful to cattle, but when moderately used as a salad green or steamed like spinach it is both tasty and nutritious.

Sweet Fennel tastes slightly licorice like anise and its oil is even used for liqueurs in certain candies. The plant was mentioned by Charlemagne and prized in Italy where it is known as *finnochio*. Sweet fennel looks like stocky celery and can be used as an appetizer, salad, or cooked vegetable. The leaves and stalks are eaten. Fennel seeds are used as seasoning for many dishes, particularly fish, and in pies. Many other herb leaves are used, in moderation, as vegetables. These include marjoram, thyme, rosemary, dill, sweet cicely, lovage, borage, chervil, sage, and mint.

Some additional edible roots are:

Arrowroot is grown primarily on Saint Vincent Island in the West Indies. A starch is derived from it, usually in fine powder form, that is used as a natural thickening.

Cassava is grown mostly in Africa, where it often replaces the yam and sweet potato as a staple. Tapioca is made from its roots. Cassava was originally an American root. It is extremely important in tropical countries because it is very hardy and will last a long time if left in the ground. When eaten to the exclusion of other foods, there is a possibility of malnutrition, because

- it is high in starch and low in protein. Manioc is a meal made from cassava root. Cassava must always be cooked; it is usually boiled and mashed, made into meal, or fermented for alcoholic drinks. Sometimes the leaves are eaten as greens.

Taro is a root that grows wild in tropical climates and is cultivated extensively in Japan, Hawaii, and the West Indies. It is a staple of people living in the Pacific region, where it is eaten like potatoes—boiled or steamed or roasted and ground into flour. Also known as dasheen, taro is the well-known poi of Hawaii. Highly digestible, the root is boiled down and fed to infants instead of milk. It is more nutritious when cooked in its skin.

Pear

Fruits and Nuts

The aristocratic quality of tree fruits is inherent in the very power of this form of plant life. . . . It twists its rootlets into the very seams of life . . . it draws water as if from a deep well. . . .

Henry Bailey Stevens

The fruit and nut trees were here first. From our earliest days on the planet, they were our means of shelter, fuel, and food. Botanically, fruit is defined as a ripened ovary consisting of one or more seeds and surrounding tissue. But the definition hardly conveys their magic and mystery; the name *fruit* even means "to enjoy." The fruit tree was there for man to dance around in celebration and in thankfulness, to honor life and death. In fruit, nature has provided a perfect cleanser, just the right amount of natural sugar for energy, and pure distilled water—all neatly wrapped in a protective covering that can be peeled off and thrown away. Most of the fruits and nuts we eat today are the result of five thousand years of cultivation. Although perishable in its raw state, a harvest can be made to last indefinitely by sun-drying, an early discovery that man has continuously found advantageous. Broadly classified as acid, subacid, or sweet, fruit falls into four general botanical categories: pomes—those with cores and small seeds, like apples; drupes, with single stones or pits, like peaches; berries, which have many seeds throughout the flesh, like grapes; and aggregate fruits, grown in clusters, like raspberries. Some fruits (like avocados) included in this book are usually not thought of as fruits at all, but as vegetables.

Almost all fresh fruits are eaten raw (best when just ripe), but are extremely versatile and can be cooked, preserved, juiced, and used in limitless ways (see Glossary, page 228). Most fruit is grown on a large scale, but a home garden can be graced with a pear or apple tree, with espaliers scaling a wall and berries growing in wooden barrels; and a dwarf citrus or peach tree adds fragrant beauty to the home.

Nuts are close relatives of fruit and have been their companions since antiquity as symbols and as food. These naturally concentrated capsules of protein and fat are sources of high energy, particularly essential to vegetarians, human and animal. Nuts carry with them the nostalgia of autumn, but can be enjoyed year round. The oldest method of cooking, roasting (probably in hot sand), is still one of the most popular; but there is much more that can be done with them (see Glossary, page 261).

APPLES

What we have come to think of as Eve's apple from the tree of knowledge was in all probability not an apple at all. The Bible merely refers to it as a "fruit" and other fruits lay claim to this dishonor. If it were an apple, it would probably have been a crab apple, because apples did not grow in that part of the world. A "golden apple" was Aphrodite's prize in Greek mythology, and although, again, another fruit may have been more apt, the expression "apple of discord" came into our vocabulary as indirectly causing the fall of Troy. It was an apple that tantalized Tantalus; and the goddess Idun in the Norse legend supplied apples to all the gods so that they could stay eternally young. The apple tree supplanted the Maypole in England as a fertility symbol, and young people danced around it after sprinkling it with cider. There was an ancient custom of wassailing trees on New Year's Day in the belief that until trees were toasted they would not bear fruit. Everyone from William Tell to Isaac Newton to Cézanne has been enamored of the apple; it has been a favorite subject for artists, musicians, poets, and writers.

Apples spread from southwest Asia to Europe. Small, sour crab apples were hoarded in caves by Stone Age men in Europe. Remains indicate that they not only ate them fresh, but preserved them by drying. These crab apples were the parents of the cultivated apple, and cultivation of this fruit began with the beginning of agriculture in Europe. Apples were central and northern Europe's most important cultivated fruit crop at the time of the discovery of America. Early in the settlement of this country apple trees were planted by the Pilgrims on an island in Boston harbor, by the Dutch in New York, by Jesuit missionaries passing through the valley of the Saint Lawrence River. Roger Williams, who founded Rhode Island, was the subject of the following story. In order to build a monument on the site of his grave, the ground was dug up many years after his death in 1683. Curiously neither coffin nor body could be found. Nearby stood an apple tree that had sent its roots down and absorbed the coffin and its contents. On close examination the roots showed a striking resemblance to the human form: a remarkable example of natural recycling! An itinerant missionary, John Chapman, became the center of another legend. He traveled through the frontier wilderness with a bag of apple seeds and copies of the writings of Swedenborg. He planted the seeds in the clearings and tore pages out of the books for those who could read. Although apple trees are now more often grown by grafting (because the process enables production of new species), there is no question of the contribution of "Johnny Appleseed." Louisa May Alcott, whose uncle was one of the early nutritional reformers, lived in a home named Apple Slump in honor of the family's favorite dessert and had her heroine, Jo, in *Little Women* retreat to the attic and munch an apple every time she needed to think.

There are more than nine hundred varieties of apples, and over 100 million bushels are grown in the United States each year. The science of apple growing is known as pomology. The apple's popularity is due to its versatility as a raw or cooked fruit and its hardiness and ability to store well. The expression "as American as apple pie" is true, but unfortunately, according to some nutritionists, indigestible (apples and pastry do not combine well). There does appear to be some validity in the slogan "an apple a day keeps the doctor away." Some apples contain a large amount of vitamin C, usually in or right under the skin. Infants are given apple juice in lieu of orange juice because it is easier to assimilate. For years, physicians treated infantile diarrhea with a diet of apple pulp only. The converse is also true; apples are excellent as a laxative. For the teeth, apples not only have a fine detergent action but, because they must be chewed well, also massage the gums. Apple pectin has been found to be a germicide. A medical specialist in Indiana found that sores kept wet with a solution of pectin healed with greater speed.

Apple cider—hard and soft—is a popular drink here and became very important in England when they were warring with France and no wine was available. Apple brandy (calvados) and applejack are more potent by-products.

A young apple tree requires from five to ten years of attentive care and growth before it is old enough to bear fruit. For backyard gardening, espaliers are preferable. They can be trained to grow against a wall and are most decorative while you are waiting for them to bear fruit.

Many expressions have come into the language using the apple. In the Bible (Deuteronomy 32:10) it says: ". . . he led him about, he instructed him, he kept him as the apple of his eye." This reference is possibly to the fact that the ancients supposed the pupil of the eye to be a round solid ball. In modern times, it is usually applied to a favorite; an "apple-polisher" curries favor. "Apple-pie order" is probably a corruption of the French *nappes pliées,* which means "folded linen" and can also refer to beds made like apple pies. "Applecart" was old slang for the human body, so it's not difficult to see how "upset the applecart" came into use. "Adam's apple" was probably derived from the myth that a piece of the forbidden fruit stuck in his throat.

Crab apples still exist, wild and cultivated. They are much smaller and more acid than standard apples and are more often used for preserves and as decorative plants.

APRICOTS

The name *apricot* derives from the Latin *praecoquum,* which means, literally, "precocious; early ripening" and probably is an example of folk etymology, because the apricot ripens before its similarly colored stone-fruit cousin the peach. The Arabs (*alburqūq*), Italians, Spanish, and French had a hand in changing its name to what it is today. It first appeared in English as *abricock* and Shakespeare spelled it *apricock.*

In ancient times, because apples were probably of very poor quality, one can assume that the "apples of gold" referred to in the Bible were really apricots, as were the fruit of which Solomon said, "Comfort me with apples: for I am sick. . . ."

The apricot grows spontaneously in its native Asia. Alexander the Great is said to have brought it to Greece around the fourth century B.C.; from there it was carried to Italy and mentioned as a Roman fruit in the time of Christ. It spread north slowly and through one of Henry VIII's busy gardeners reached England. It came to the United States around 1720, when the Spaniards brought it to Mexico and then to California. Modern varieties of the apricot are not so different from those cultivated two and three thousand years ago. Apricots now grow throughout the warm temperate zone, mostly in China, Japan, North Africa, and California.

Apricots are cooling and stimulating and may have some remedial value to the kidneys and bladder when eaten with cream. Dried apricots are much higher in protein and vitamin A than fresh ones, but it takes six pounds of fresh apricots to produce one pound of dried, so naturally there is a big difference in cost, too.

AVOCADOS

The avocado is no relation to the pear despite the misnomer alligator pear. Perhaps this name came about because of a fairly farfetched notion that the fruit was shaped a bit like a pear (although it is just as often round) and has corrugated skin like an alligator's (although it is just as often smooth). Avocados are native Americans, originally grown in Mexico and Central America. Early Aztec pictographs had a sign for avocados, and the Aztec leader Montezuma presumably served them to Cortes when he arrived in Tenochtitlan (now Mexico City) in 1519. The early Spanish spelling of the name was *ahuacatl,* but many modifications took place and

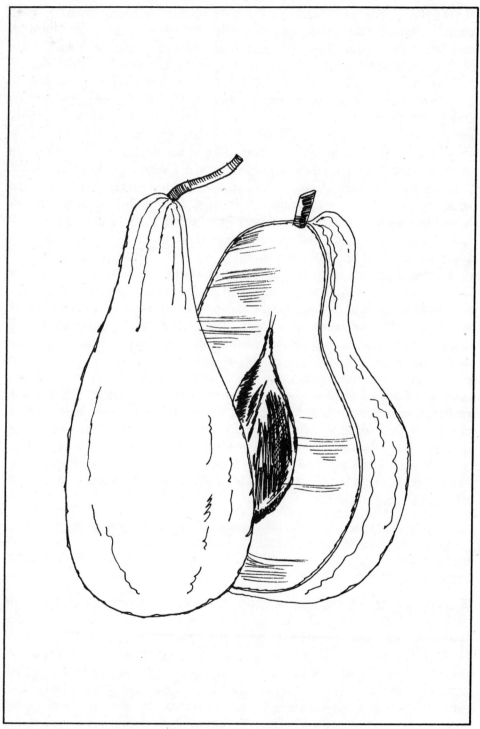
Avocado is chemically more like a nut and eaten like a vegetable.

the English spelling is closest to the later Spanish *bocado,* meaning "delicacy." Avocados do not propagate easily and planting in other countries was late. Hawaii grew them starting around the 1820s, but the industry now in California and Florida did not start until the beginning of this century.

There are two species of avocados: thin- and thick-skinned, ranging in color from purple-black to green, and both are available here. An average tree yields about five hundred fruits but can produce as many as two thousand. If you live in a place where the weather isn't really hot most of the year, don't expect to grow a tree that will yield fruit. But if you like to eat avocados and have a window, you can become a showy gardener simply by planting the pit, base down, in some soil. Wait patiently and when the stalk grows to about six inches, be ruthless in pruning. Avocado plants have been known to reach mammoth proportion indoors and are beautifully decorative. I don't ascribe to the pit-and-toothpicks-in-water method. It's much easier to plant directly in soil.

Avocados are possibly the highest-evolved of all nutritive plants, the fruit taking nine months to mature, as does the human embryo. They contain up to 30 percent of their weight in oil, which is the largest percentage of any fruit (with the exception of olives), and can supply the need for oil in diets where other means are not permissible. They are also extremely high in protein and chemically are more like a nut than a fruit. Their protein content makes them an important meat substitute. In Mexico and Central America, tortillas, avocado, and coffee are considered an excellent meal. In Brazil, avocados are used more as a dessert and are often made into ice cream. In this country, avocados move easily throughout the courses of the meal (see Glossary). They combine well with either fruits or vegetables and are easily digested and a thrill to the senses. An excellent greenish-colored oil is extracted from avocados, which is used as a salad oil and as a base for many cosmetics and shampoos.

BANANAS

Bananas are grown on a giant herb, probably the largest plant on earth without a woody stem. The Koran took liberties by referring to it as the "paradise tree," and a medieval legend even claimed it (and not the apple) as the forbidden fruit. Its earliest beginnings were in the humid tropics of southern Asia. Early man discovered that roots could be dried and carried great distances to be grown under suitable climatic conditions. During the great migrations from southern Asia, the banana became a traveler and found root in the Pacific Islands. The armies of Alexander the Great found bananas growing abundantly in the valley of the Indus in 327 B.C. It is told that the sages of India rested in the shade of banana plants and refreshed themselves with the fruit, so that the banana's name became *Musa sapientum,* which means "fruit of the wise men." Legends often have origins in fancy rather than truth; there are many trees whose shade is far better than the banana's and other fruits that have claimed this honor (see *Mangoes*). The Arabs found bananas in India and probably introduced them into the Holy Land and northern Egypt. They were carried by them from tribe to tribe across equatorial Africa to the coast of Guinea along with their other cargoes of ivory and slaves. The Portuguese found them there in 1492, bearing the African name (*banana*), and brought them to the Canary Islands. They were brought from the Canaries to Santo Domingo by a Spanish priest, Friar Tomas de Berlanga, in 1516. In the mid-nineteenth century, people were writing of bananas as the delicious fruit of the tropics. By 1850 ships were bringing cargoes of them to the United States. In 1876 bananas wrapped in tinfoil sold for ten cents each as a novelty at the Philadelphia Centennial Exposition. Captain Lorenzo Dow Baker (and partners) founded the United Fruit Company in 1899. This supposedly came about because he was a man poorly affected by cold weather and he sailed to the Caribbean to follow the sun. There he found bananas growing in Jamaica, took a great interest in them, and experimented with keeping them in a dark corner so they would not rot. His success led to the huge export company he founded, and to a whole new trade. Certain countries on the mainland of South and Central America became known as banana republics, with many social and political reverberations.

Cultivated bananas today derive from the hybridization of two wild species which still exist, but which have sterile, seedless fruits and so must be continuously planted. The stalk of the banana plant is made up of clusters called hands. Each hand has from ten to twenty bananas, called fingers. The word *bunch,* although descriptive, is incorrect. In one thousand square feet of space it is possible to produce four thousand pounds of bananas. In the same space one can produce thirty-three pounds of wheat or ninety-nine pounds of potatoes. It is understandable that the banana is the most important tropical fruit and a staple in most diets. Many natives living outside of the cities in the tropics need only a few plants in their garden for table use. Lesser-known types are small lady's finger and red-skinned bananas. Bananas are even used for making beer in East Africa.

The banana ranks very high nutritionally. It is richer in minerals than any other soft fruit except strawberries; it has a high caloric value (but can also be used successfully in reducing diets); it is a natural laxative, a soothing addition to the diet of the ulcer or colitis patient, and an aid in the treatment of kidney ailments. For a long time people with digestive difficulties were warned against eating bananas, and they were absolutely prohibited in the infant's diet. This was probably because green or underripe bananas have a heavy starch content, which is difficult to digest but which changes completely to digestible fruit sugar when the fruit ripens. Today bananas are usually among the first solid foods given to infants and are standard in the diet for celiac disease, which affects children. Though I recommend no such extremism, it is interesting to note that the pygmies of Central Africa eat as many as sixty bananas a day.

Some slang terms using banana are "top banana," usually referring to the headliner in a vaudeville act, and "to go bananas," which probably has something to do with acting like a monkey. The term for an Australian one-pound note is just plain "banana."

BARBERRIES (see Wild Berries)

BILBERRIES (see Blueberries)

BLACKBERRIES

Blackberries have grown wild in the temperate regions of the world from earliest times but are not cultivated on a large commercial basis. They appear in many legends. Possibly Moses' burning bush was a blackberry bush; Christ's crown of thorns may have come from one and perhaps that's why it is said that the devil hates blackberries.

Blackberry cordial is a delicious sweet drink, but if you want a medicine, an infusion made from the roots or leaves is considered valuable in the treatment of diarrhea. The leaves have been used for such diverse problems as loose teeth, snakebite, rheumatism, and popeyes. For burns, try applying blackberry leaves wet with spring water.

Dewberries are a more delicately flavored type of blackberry grown on trailing rather than upright plants.

Loganberries are dull red in color and more acid. They are grown mostly for canning or making into drinks.

Lesser-known blackberry relatives are cloudberries, which fruit in northern regions such as arctic Russia and within the Arctic Circle; and boysenberries, which add their flavor to a delicious sherbet and yogurt and have been praised in the Simon and Garfunkel song "Punky's Dilemma."

BLUEBERRIES

> "I'm as normal as blueberry pie."
> Rodgers and Hammerstein

Blueberries are native Americans. In 1616 Samuel de Champlain found the Lake Huron Indians gathering blueberries. He wrote, "After drying the berries in the sun, the Indians beat them into a powder and added this powder to parched meal to make a dish called 'sautauthig.' " In the northwest territory, Lewis and Clark saw Indians smoke-drying the berries for winter use in soups and stews and were served venison cured by being pounded with blueberries and then smoked.

Blueberries can be high-bush or low-bush, wild or cultivated. In the United States, both kinds are sold commercially. The wild blueberry capital is Maine, and the industry prides itself on the first commercial use of blueberries in the world: canned for the Union troops during the Civil War. More cultivated berries are produced in the United States, but bakers prefer the wild ones for their small size and flavor. Blueberries are only produced on a large commercial scale in the United States. Elizabeth White, living in New Jersey around 1900, seems to have been responsible for beginning the first commercial cultivated crop. She offered prizes for bushes bearing the largest blueberries. Then she began cultivating and cross-breeding.

A sibling rivalry seems to exist between blueberry enthusiasts and huckleberry enthusiasts. The former claim that the other berry has ten large bony seeds that do not disappear when you eat them (as do the soft ones of blueberries) but instead can get stuck in your teeth and become very unpleasant. Some renegades have gone so far as to say that if Mark Twain had tasted the "real" thing he would have named Tom Sawyer's friend Blueberry Finn—and who knows what effect that might have had on American letters!

In addition to the huckleberry, the blueberry has a few unlikely relatives such as the cranberry, the mountain laurel, and the azalea. Blueberries had a medical use. A Scotch medical book reported: "Fluxes are cured now and then by taking a spoonful of the syrup of blueberries." Russian women used to prescribe a preparation of dried blueberries called *chernika* for bellyaches, and today we have blueberry soap, which smells delicious. Baking hint: when using blueberries in cakes, muffins, etc., dust them with flour before combining with batter and they won't sink to the bottom (this also applies to raisins).

Bilberries are related to blueberries and are native to Britain, parts of continental Europe, and northern Asia. Sometimes they are called whortleberries. Bilberries are not eaten raw so much as dried for use as treatment for diarrhea. They were once used during typhoid epidemics in northern countries.

Cowberries are another relative of blueberries. They grow in northern climes. Because they are red, they are sometimes known as mountain cranberries.

CHERRIES

Cherry pits have been found in Stone Age Swiss lake dwellings, ancient Scandinavian deposits, and cliff caves of prehistoric inhabitants of America. The modern cherry probably originated in China over four thousand years ago and was brought to Rome from Kerasos, a city in Asia Minor that gave its name to the fruit. It was later brought to England, but it was not until Henry VIII's gardener planted cherries in Kent that they became popular. The Montmorency, the foremost tart variety, is grown there. Cherry seeds were brought to the United States on the *Mayflower;* in the West, Spanish missions began cultivating large orchards, and Oregon's huge sweet-cherry industry was begun in the mid-nineteenth century.

Cherries fall into two categories: sweet and sour. Sour cherries can grow in temperate climates anywhere in the world and cultivated varieties number about three hundred. The best-known sour cherry is probably the dark morello. Sweet cherries must have a temperate climate to grow; they number about six hundred varieties, are usually larger than the sour, and the Queen Anne and Bing are among the most popular types. The growth of a cherry is a marvelous example of self-sufficiency. The pit hardens and becomes a stone as the fruit ripens. As though to make up for the pit's contracting, the fruit fills out around the pit until it is plump and juicy at full ripeness. Wild cherries multiply rapidly and haphazardly, with the help of birds who distribute the seeds. But birds also become a serious problem in cultivated orchards during harvesting because they often damage as many cherries as they eat. In large orchards the practice of using firecrackers has been successful, for the birds never seem to accustom themselves to the noise and fly away after each timed explosion.

> Those cherries fairly do enclose
> Of orient pearl a double row,
> Which when her lovely laughter shows,
> They look like rosebuds fill'd with snow.
>
> Thomas Campion, "Cherry-Ripe"

In areas where cherry blossoms abound, spring is a special delight. The Japanese cherry has little food value, its fruit being too small and acid, but Japan is alive with cherry groves because of the color and form of the blossoms. April is their cherry month and the orchards become festive sites for great conviviality. A fifth-century Japanese emperor was sailing on a lake beneath some cherry trees and a few pink blossoms floated down into his sake cup. He was so struck by their beauty that he took to drinking his wine beneath the trees every spring season, and to this day, wine drinking is very much a part of the celebration. There is a story that a Japanese warrior had for many years in his youth played beneath the branches of a cherry tree at Iyo. He reached a great age, outliving all his family and friends, and the only object that linked him to his past was his beloved cherry tree. One summer it died. He took this for a sign and was not consoled when a young sapling was planted nearby. That winter, he spoke to the dead tree and pleaded with it to bear blossoms just once more, promising that if it did he would give up his life. Then he spread a white sheet upon the ground and committed hara-kiri. As his blood soaked into the roots of the tree and his spirit into its sap, the dead tree burst into bloom; and every year on his death day, though the ground is hard and all other trees dormant for the winter, this tree at Iyo blossoms.

We have no such splendid legends in this country. In fact, though I risk being called unpatriotic, the only truth we have of George Washington's "truthfulness" was that his father was a farmer who valued his cherry trees, and it gives us an occasion to celebrate his birthday with traditional cherry pie. A cherry tree did influence the course of New York City's Broadway. At the corner of Broadway and Tenth Street, where Grace Church now stands, Hendrick Brevoort had an inn with a beautiful cherry tree in the yard. To spare the tree, the compassionate city planners diverted Broadway, and so the street becomes a diagonal thoroughfare, cutting slowly to the west as it goes north.

Often cherry wood is used for beautiful furniture, but the bark is employed sometimes in folk medicine as an astringent. There seems to be evidence that cherries and cherry juice have a positive effect on gout and arthritis, possibly because of the substance contained in the pigments. The old warning of "Don't drink water when eating cherries!" appears to have some scientific validity. The kernel proteins of the cherry can become poisonous when combined with water. Also, the skin of the fruit sometimes interferes with sensitive stomachs, and in those cases it is advisable to sieve cherries and drink the juice.

The cherry has figured prominently in slang. Perhaps the most common use is the reference to a young girl, usually a virgin. Any number of explanations come to mind but, in the interest of propriety, I leave these to the reader's imagination. "Cherry-merry" and "cherry-nose" both refer to states of drunkenness and, for no apparent reason, inferior seamen were known as "cherry-pickers" during the nineteenth century.

CITRONS (see Miscellaneous Citrus Fruits)

COCONUTS

There is a South Sea proverb that says, "He who plants a coconut tree plants food and drink, vessels and clothing, a habitation for himself and a heritage for his children." And it is true, for a coconut can be a "tree that provides all the necessities of life" (as its Sanskrit name *kalpa vrisksha* means). The trunk is used for building material; the leaves can be used for paper stock, baskets, or hats; the leaf stalks can fence in a garden; the coir (fiber) can be made into nets, ropes, brushes, and mats; the meat and milk for food; the empty shells for bowls and spoons, and the hollowed-out trunk for a canoe.

Coconuts are native to Malaysia, southern Asia, and tropical South America. They were carried in prehistoric days across the Pacific to their favorite home, the South Sea Islands. Early European explorers found them growing in America, probably having floated over the sea at an earlier date. Today they grow in the West Indies, Ceylon, India, Puerto Rico, Florida, and tropical coasts all over the world.

A coconut palm can grow from sixty to one hundred feet and matures in seven years. There can be as many as two hundred nuts a year on a tree. The nuts grow shielded from damage by the fibrous husk. All of the coconut inside is a seed, the largest seed known to man. When the nut is still green, the inside is a white creamy substance, which one can eat with a spoon, and a sweetish water (the milk). When it ripens, the outside skin

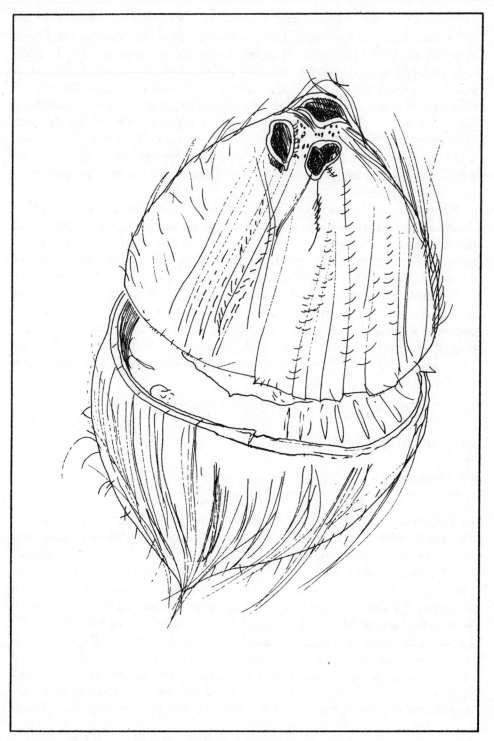
Coconut milk can be obtained by piercing the soft "eyes."

becomes brownish, harder, and woody, and the coconut meat (copra) is formed. There is less milk. There are many food elements in the brown skin that clings tightly to the meat, so this should be eaten as well. In addition to the food it yields, the coconut also supplies an oil which is used in margarine, medicines, soap, and cosmetics and which is particularly valuable as a tanning agent. The sap of the coconut trunk, upon evaporation, yields a crude sugar. The best way to eat a coconut is to pierce the eyes (the three soft spots at the end of the shell through which the young shoot and root emerge at germination). Drain out the milk and then tap the nut all over with a hammer until the shell cracks, or heat in a medium oven for about a half hour.

When nuts fall from the tree and crack open, they can take root in about three years and form a new palm tree if allowed to remain on the ground in the sun. A more modest home palm can be yours by planting a whole coconut in water and then soil, making sure it stays warm. Don't expect either flowers or fruit from it—only beautiful leaves.

Coconuts are a staple in tropical diets; the milk often replaces cow's milk in India. It is estimated that coconuts are the only source of fat for over 400 million people in tropical countries, and the annual world production of coconuts is nearly 2 billion. A somber note for coconut lovers: the palms in Florida have been dying from a mysterious disease and three thousand were lost during 1972.

COWBERRIES (see Blueberries)

CRANBERRIES

Cranberries, although married to turkeys during Thanksgiving, are actually much older than that holiday. They are one of this country's few truly native fruits and were used by the Indians long before the Pilgrims arrived. The Indians crushed them and added them to meat (a natural preservative), used them in poultices to draw poison out of arrow wounds, and made red dye out of them. Perhaps because the Indian names for the berry were unpronounceable, none of them were adopted by European settlers. Instead, the name is a contraction of *crane berry,* because the pink blossoms resembled the head of a crane, and possibly because cranes could often be seen eating ripe berries. Cranberries grow on swampland (bogs) in New Jersey and Cape Cod. American sailors used cranberries as a preventive for scurvy in the same way their British counterparts used limes.

The first commercial use of cranberries was in 1830, and cultivation moved westward so that now they are also produced in Wisconsin, Washington, and Oregon. In 1959 the industry had some unfavorable publicity (sometimes referred to as "the great cranberry crisis") when a chemical weedkiller, capable of producing cancer cells in mice, was found being used in cranberry bogs. The sale of cranberries was temporarily suspended. Now the industry is larger than ever and cranberries are popularly eaten in sauce or relishes, and drunk as a juice, alone or in combination with other fruit juices. Cranberries are one of the three fruits causing an acid reaction in the body. A cranberry must "bounce" to prove it is fresh; otherwise it is discarded. Cranberry juice is used as a treatment for cystitis.

CURRANTS

Fresh currants are berries, in the same family as gooseberries. They are black, red, or white. (Dried currants are not currants at all but raisins made from the small seedless grapes of Corinth, hence their name [Anglo-French: *raisins de Corauntz*].) Black currants grow mostly in northern Europe and Asia. They are the basis for the excellent liqueur called cassis, and their leaves have been used to treat throat infections and coughs. Red currants are more common in the United States and have many uses, as a fresh fruit and in jellies, sauces, baking goods, and drinks. White are really red without the pigment.

DATES

There seems to be some substantiation that date palms grew in the Garden of Eden, largely because of the area where that paradise was supposed to have existed. The date is probably the oldest cultivated fruit. The Babylonians are believed to have cultivated dates eight thousand years ago; and they have been found in early Egyptian tombs. "Honor your material aunt, the Palm," said Mohammed, "for it was created from the clay left over after the creation of Adam." And from ancient times the date palm has been revered as a symbol of beauty, majesty, and triumph. Its scientific name is *Phoenix dactylifera,* and like the mythological phoenix, rising from the ashes, the tree has its feet in water and its head in the fires of heaven (irrigation provides the former, and the desert sun, the latter). It was worshiped by the Muslims as a tree of paradise, its fruits having the same taste on earth as they have in heaven. In fact, a beverage made from fermented dates and water is not prohibited by the Koran, as are other spirits. The Jews carried branches on solemn festivals to commemorate their fathers' gaining possession of the Promised Land. The Tamanaquas of South America believe that the human race sprang from the fruits of the date palm. Two clumps of date palms planted by Alexander the Great's armies twenty-two centuries ago to mark his outposts still flourish in the Indian Himalayas.

There is a story of an Arab woman who worked as a governess in England for several years. Her reports upon returning were of a country like a garden, full of fine horses and rich people who were all wise and happy. Hearing this, the villagers were envious and desired to travel there until she said, "One thing is certainly missing—there is not a single date tree in the whole country." That was enough to convince her listeners that their country was better. England, of course, like other temperate countries, is not suited to growing the date palm. It thrives in dry subtropical areas such as Morocco, Saudi Arabia, Algeria, and Egypt. Iraq is the most important date-producing country, but dates are also successfully grown in the drier regions of the United States, such as California and Arizona. Although the industry here is small—less than 40 million pounds out of a world production of 4 billion—dates packed in California go to some sixty other countries and supply most of Europe. The California date industry began only about fifty years ago as a means of using desert lands; and the Coachella Valley successfully produces the deglet noor, meaning "fingers of gold" (Algerian), and medjool (Moroccan) date trees, which are being threatened with extinction from disease in their native countries.

Date palms have often been compared to humans. The tree is covered with fiberlike hair and if leaves are cut off, unlike many other trees, none will grow in the same place (analogous to man's inability to grow new limbs). There are also two sexes. Male and female palms grow apart and an orchard is much like a harem—usually one male to fifty females. But the trees have a longer life span than humans do, fruiting up to the age of eighty years. Pollination must be done by hand; the blossoms do not attract bees. A good tree can yield two hundred pounds of fruit a year, the fruit protected under paper "umbrellas" during the ripening process. In addition to its rich fruit, rope can be made from the fiber, baskets from the cane, and building materials and thatch from the leaves. Roasted pits have been used as a coffee substitute; a water-clouding drink, similar to anise, called arrack is made from the sap; and, finally, when the tree is too old for anything else, the trunk can be used for fuel. It is no wonder that there was an expression that the desert Arab would part with his donkey, camel, or even one of his wives if offered a profitable exchange, but if asked to sell a fertile date palm he would reach instantly for his dagger.

There are many different varieties of dates but they are classified generally into three types: (1) fresh and soft—often sold in a pressed mass and rarely seen in the United States; (2) semidry (such as deglet noor)—one of the most popular, often sold with the fruit still on the strand; (3) dry—quite hard and usually ground into flour. A camel can live on dates and water alone, but for Westerners the date is an excellent confection, healthier than candy and, according to the Department of Agriculture, less fattening than even a low-calorie rye wafer. For the dieters, let them eat dates (as long as they can resist temptation and eat only two!).

DEWBERRIES (see Blackberries)

ELDERBERRIES (see Wild Berries)

FIGS

The fig is one of the most ancient fruits known to man. Fossil figs have been found in deposits in France and Italy dating from the Tertiary period (which extends from 65 to 3 million years ago). Of course, these were wild figs, but cultivation took place as early as 3000 B.C. in western Asia. There are many references to figs in the Bible and other religious writings, the most familiar being of Adam and Eve who "sewed fig leaves together and made themselves aprons." There is some dispute among scholars as to whether the tree was actually a fig, the leaves being rather thin and rough and not too suitable to sewing, but we must allow for poetic license. (In the East, fig leaves are still sewn together by the natives and used for wrapping fresh fruits sent to markets.) There is certainly no doubt that Canaan was "a land of wheat and barley, and vines, and fig trees . . ." and two hundred cakes of figs were included in the presents that Abigail offered David. In Christian mythology, Mary hid the infant Jesus from Herod's soldiers in the trunk of a fig tree, which opened for her at exactly the right time. Saint Augustine, sitting beneath a fig tree, wrestling with doubts about some statements in the Scriptures, suddenly heard the tree speak to him in a child's voice, instructing him to read again; he did and believed. Christ came upon a fig tree when he was hungry, only to find it bare. In Eastern mythology the fig tree is also associated with divinity, for the bo tree under which Gautama sat and Buddhism was born was a species of fig. The fig is prominent in classical mythology: Lyceus, a Titan, was changed into the tree by Rhea; the fig was carried beside the grapevine in processions in honor of Bacchus, and though his debauchery may have stemmed from the grape, his vigor was credited to the fig. Because a fig tree was found growing where Romulus and Remus were cradled, it was worshiped by the Romans, and women wore collars of figs to festivals. Fig wood was used by Egyptians for mummy cases and the Greeks considered it such a necessity that its exportation was prohibited. (The word *sycophant* comes from the Greek *sykon*, "fig," and *phainein*, "to show," and originally referred to one who accused fig stealers.)

The symbolism of the fig as an emblem of fertility can possibly be explained by the unique way in which it is pollinated (see next paragraph) and it appears as such often in mythology and history. One example is the way human scapegoats were punished for drought or famine. They were beaten with fig branches and strands of figs were hung around their necks (white for women, black for men), which symbolically represented the "marriage" of the trees. The whole ceremony was performed to assist the trees in their fertilization. To this day in Italy the fig has a pejorative sexual connotation in certain obscenities, both verbal and visual, and in Shakespeare's plays it is often an indecent symbol and one of mockery. Examples like "Fig me!" and "The figo for thee" need little explanation and, even today, it's a matter of some contempt to say one doesn't "give a fig."

Figs are unique in several ways. The flowers are not borne on the tree, like the peach or the apricot, but rather the fruit itself is a fleshy receptacle enclosing a multitude of flowers that never see the light but ripen the fruit from within. Pollination of many fig trees (the most famous are those grown in Smyrna where fig culture has been a principal industry for two thousand years) is done by a special process called caprification. The Capri fig tree bears a fruit that is inedible but contains within the fruit tiny wasps called *Blastophaga*. The Capri figs are picked and carried to the female Smyrna orchards, where they are hung inside of paper bags or baskets on the trees. Once there, the female wasp tries to enter the Smyrna fig to deposit her eggs. She is unable to do so, but in trying, she brushes off the essential pollen necessary to fertilize the Smyrna's inner blossoms. The male and female have no direct contact; the wasp is the only way the Smyrna fig can be pollinated—a remarkable cooperative effort of the animal and vegetable worlds. California produces this type of golden-colored fig, called Calimyrna (a compound of *California* and *Smyrna*) on a large commercial scale. Other types of figs that need not be pollinated in this fashion are other white figs such as the white Adriatic and the kadota (mostly used for canning) and the black figs such as black mission, named

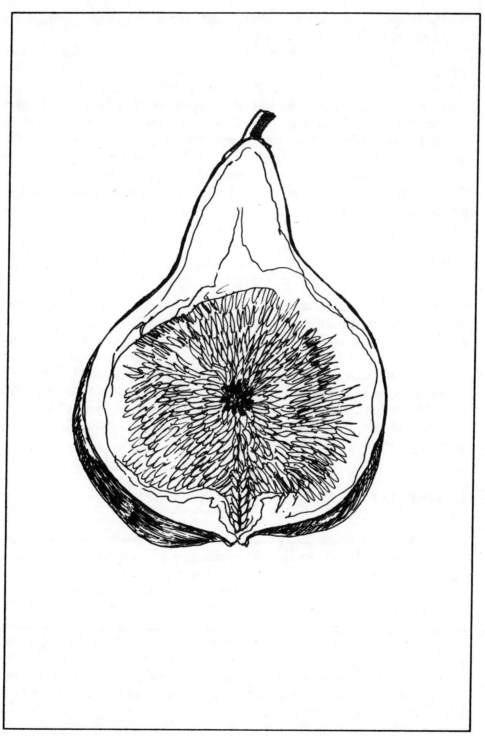

The flowers, borne inside the fig, develop into its fleshy center.

for the missions along the California coast, where it was grown originally in the mid-eighteenth century.

Although figs are eaten fresh in season (fresh figs and prosciutto is a delicious Italian appetizer), they are most often eaten in dried form. Drying figs is relatively easy, because a ripe fig falls to the ground partially dried and the sun removes the rest of the moisture. The facility and value of dried figs probably have accounted for their popularity throughout time.

"And Isaiah said, Take a lump of figs. And they took and laid it on the boil, and he recovered" (2 Kings 20:7). In addition to this use, figs are still employed medicinally in many parts of the world—cooked in milk for ulcerated gums, juiced for treatment of ringworm and warts, and valued as a laxative.

GOOSEBERRIES

Gooseberries may be reddish or dark green, but the most common ones are yellowish or whitish green. The skin is somewhat translucent so that the "veins" show through, and the berry has a slight downy texture (possibly the explanation for its name). Gooseberries and currants sometimes bear a fungus that attacks the important American white pine and so are often prohibited from being planted or shipped into areas where this tree is grown. Gooseberries have been domesticated for only about one hundred years. If you can find them, the leaves are very tasty in salads. Gooseberries are also grown in Europe.

GRAPEFRUIT

The grapefruit is descended from the pomelo or shaddock (see *Miscellaneous Citrus Fruits—Pomelos*). The fruit grows in clusters and probably received its name because of a somewhat farfetched resemblance to a bunch of grapes. The tree was first planted in the West Indies by the Spaniards, and Ponce de León is credited with bringing it to Florida in 1513. By 1830 it was recognized as a distinct species and commercial cultivation began about fifty years later. It was not until the early 1900s that it became popular throughout the world. As strange as this sounds to us today, in the mid-1800s the grapefruit was known as the "forbidden fruit," probably because it was hazardous to eat fresh owing to the long time lapse and lack of refrigeration from grower to market.

Grapefruit can be grown in the subtropics and tropics and are produced on a large scale in Florida, California, the West Indies, and Central and South America. They can have yellowish or pinkish pulp (Texas mostly produces pink grapefruit), and, as with oranges, seedless grapefruit have been developed.

Like other citrus, grapefruit has a place in home remedies (see *Lemons*), and the inner rind is a source of quinine and also may be used. Grapefruit is good for the skin, blood and treatment of jaundice; it is a good eliminator and in recent years has become very popular as a diet food. In fact, I know of some really determined types who dieted exclusively on grapefruit, although it is hardly a "complete" food. There seems to be some negative opinions on grapefruit and its effect on the teeth.

GRAPES

It was on a grapevine swing that man first teetered a little nearer the stars.
Cameron Rogers (ed.), *Full and By*

The grape is one of the oldest plants cultivated by man, and it is impossible to consider man's history without it. Fossilized leaves that could be 70 million years old have been found in the Northern Hemisphere, and they bear a startling resemblance to those of our present-day species. Grapes were the vines that Adam, Eve, and Noah planted. Saturn gave the grape to Crete, Osiris to Egypt, Geryon to Spain. Dionysos (Bacchus) was the Greek and Roman god of the vine, and those who worshiped him were notoriously addicted to wine, wild dances, and hedonistic excesses. In the Bible we read that the Israelites sent spies to report on the promised land of Canaan; they returned with the legendary bunch of eschol that needed two men to carry it. The surprised Israelites, used to small Egyptian grapes, were pleased to find such largesse in the barren and sandy desert. (Today the valley of Eschol is situated near Bethlehem.) Moses exempted vineyard planters from military service and wrote a law prohibiting the pruning and harvesting of vines in sabbatical years. There was a vine sculpted from gold and grapes made from precious stones on the eastern wall of the old temple in Jerusalem. Clusters of grapes were used on ancient coins (and some recent Italian and Israeli ones); grapes often appear as a motif on buildings and in illuminated manuscripts. In the Far East, the grape was embroidered on garments and used on crests; and in the recent past taverns were designated by a sign on which was painted the picture of a bush. The grape became one of the symbols of Christianity; Jesus said, "I am the true vine, and my father is the husbandman." Vines decorated the catacombs; wine is used in the sacrament of Communion in Christian churches. In fact, monasteries have always played an important part in grape growing. Grapes were grown by monks for religious observances and then became an item for general consumption and exportation. In California, the early missions were active grape-producers and many a California wine label bears the word *Mission*. In wine-producing countries, grapes are naturally revered. Often a bunch will be hung in doorways to announce that the new wine is ready to be tasted (warning: new wine tastes very mild and is very potent!). There is a legend that once a year the spirit of Charlemagne walks beside the Rhine, and then to the center of the river. If he is seen to lift his hand in blessing, a good vintage will follow.

Grape growing is called viticulture (the botanical name for the vine is genus *Vitis*). Production is wholly in temperate and subtropical regions. The Old World grapes probably originated near the Caspian Sea. These require warm, dry climates and are more sensitive to changes in temperature than the American types. Growing is done by grafting, rather than seed, and grapes are grown mostly on terraced land or arbors and trellises. A grapevine is an object of beauty. It is gentle and must always be supported. Often vines are planted around rows of trees, the trunks and branches serving as their support. The United States produces more grapes than any other fruit, with the exception of apples and oranges, and California is responsible for 90 percent of the entire country's production. Grapes are red, black, green, or white; seedless or seeded. Some grapes are grown strictly for wine, champagne, or brandy; others for dessert; others for drying into raisins.

The grape's good blending of acid, natural sugar, mucilage, and bitter and astringent properties makes the fruit acceptable to even the most delicate stomach, and the grape cure (where one lives solely on grapes and grape juice for several days or several weeks) is employed often as a temporary purifying or reducing diet. In some European spas, particularly those along the Rhine, people with impaired health are allowed to eat only ripe grapes and a little bread. Grape poultices have been used for intestinal fevers. When grapes are used as a treatment, it is preferable to avoid all other fruits. If grapes are allowed to ferment, the "cure" is quite different! Fermentation is the process by which yeast converts the sugar in the grapes into alcohol and carbon dioxide. In making wine, the gas is allowed to escape, but for champagne and sparkling wines, it is retained and forms the characteristic bubbles. Often the name of a wine is derived directly from the name of the grape, e.g., muscat to muscatel. "Grape fever" is a term applied to that bug which causes men to plant or invest in vineyards.

The word *grapple* probably derives from the way in which grapes are gathered by pulling with hooks. "Grapevine" is the intangible and untraceable means by which rumors are spread; grapes are the rumors—an easily understood analogy when one observes the way vines grow. The expression "sour grapes" originated with Aesop's fox who, upon realizing that he couldn't reach the grapes, justified himself by saying they were sour, but Uncle Remus said, "Winter grape sour, whedder you kin reach 'im or not." In the Old Testament (Jeremiah 31:29) it is written: "The fathers have eaten a sour grape, and the children's teeth are set on edge," and there are, of course, the "grapes of wrath."

Raisins   People probably discovered raisins accidentally by allowing grapes to ripen too long on the vine. Soon they became sun-dried, and raisins became a part of the diet. This happened at least four thousand years ago. By 1000 B.C. the Israelites were using raisins to pay their taxes to King David. Six hundred years later, Armenia became the center of a large raisin industry that was to last for the next fifteen hundred years. Today the vineyards of the San Joaquin Valley in California produce more raisins than the rest of the world and many of the producers there are of Armenian descent. The word *raisin* is probably a corruption of the Latin *racemus*, which means "a bunch of grapes." Raisins can also be fermented into a wine. They are eaten or cooked like other dried fruits. Sultanas are small, seedless raisins, and dried currants are even smaller and not to be confused with fresh currants (see *Currants*).

HUCKLEBERRIES

The huckleberry is reputedly the oldest living thing on earth. One plant found in western Pennsylvania covers several square miles and is estimated by botanists to be over 13,000 years old (older than the oldest California redwood). It is one of the last surviving examples of the box huckleberry. Huckleberries have harder seeds and are smaller and more tart than blueberries. They are not produced much commercially, can be found growing wild in woods and along lakeshores, and are used generally in baking and preserves (see *Blueberries*).

JUNIPER BERRIES (see Wild Berries)

KUMQUATS (see Miscellaneous Citrus Fruits)

LEMONS

Lemons, like other citrus fruits, probably originated in southeast Asia. At one time the lemon capital of Europe was Messina, famous for its thin-skinned, fragrant lemons. Today they are widely grown in subtropical countries, throughout the Mediterranean region, and in the United States, mostly in California and Florida. The lemon's shape is supposed to be distinctive because of the blunt point at the apex, but I've often seen perfectly round ones. Lemons may be harvested at any season and often flowers are found on the same branch as mature fruit. There are sweet lemons grown in some oriental countries but rarely seen in the United States.

Acid (one ounce of lemon juice contains approximately forty grains of citric acid) predominates in lemons. The whole lemon, including the rind, is used in food and drink preparations and even as a tenderizer (see Glossary). But its use doesn't stop there. There are many commercial preparations that include lemon, and the list of home remedies is extensive. Some of the illnesses and complaints you can treat with lemon are the following: (1) Colds and coughs: drink juice "straight" or mix with syrup or another juice. (2) Fevers: lemon is a powerful diaphoretic (in simpler language, a promoter of perspiration). In fact, heating lemons helps to form salicylic acid, one of the main ingredients of aspirin. (3) Indigestion and other stomach disorders: juice of half a lemon combined with a quantity of salt and taken immediately after meals. (4) Reducing diets: barley water flavored with lemon rind; or juice with water (lemon contains a reducing sugar that helps keep weight down). (5) Skin and scalp problems: mixed with oil for a good skin "food"; rubbed into scalp for dandruff and falling hair. Many commercial cosmetics and shampoos include lemon oil both for perfume and medicinal use. (6) Superficial wounds: dried lemon powder sprinkled on wounds is an effective clotting agent. (7) Warts and corns: for the former, the rind is steeped in vinegar and then rubbed on; for the latter, juice is rubbed in. (8) Liver, kidneys, and bladder: juice in the morning. (9) Rheumatism, gout, neuritis: drink undiluted juice with a pinch of powdered garlic on an empty stomach. (10) Bleaching stains, relieving sunburn, chapped lips: applied with glycerin.

Some home-remedy "experts" advise the following method for extracting lemon juice. Bake lemons in a moderate oven to heat thoroughly and then roll them in the palm of your hand to loosen the juice. Lemonized milk may be made from the juice of two baked lemons and one pint of milk, with a dash of nutmeg or mace. The latter is supposed to be particularly effective in treating anemia. From the foregoing, one might conclude that lemons are extremely healthful. There is some evidence to the contrary in the area of the teeth. This evidence (which also includes the other citrus fruits) claims that these fruits, and particularly juices derived from them, when consumed in excess, may have an erosive action on the teeth and gums. The best policy seems to be one of moderation, with emphasis on eating the whole fruit rather than just the juice.

Lemon seeds, like other citrus seeds, are quite easy to pot indoors. They can be placed directly in soil and should be kept warm but can survive in varying conditions of light. They should be watered frequently. If one is patient, a tree may bloom after about five years and produce ripe fruit about a year later. But even without flowers or fruit it is still a handsome tree in any apartment or house.

There is an Eastern myth that all the ogres dwelling in Ceylon live in a single lemon and if one can find the lemon and cut it in pieces, they will all perish. It is no myth, unfortunately, that there are many automobiles unaffectionately referred to by their owners as "lemons."

LIMES

Many people associate limes with British ships (lime-juicers) and sailors (limeys) because lime juice used to be served on these ships to prevent scurvy, a disease caused by lack of vitamin C. Ironically, lime juice has less vitamin C than lemons, oranges, or grapefruits. But, then, imagine calling anyone a "lemony"!

Limes are less hardy than the other acid citrus fruits and are grown mainly in tropical countries, often replacing the lemon in use there. In fact, because the tree is so small, it is often grown in private gardens, where the fruit is easily accessible to the kitchen. Limes are usually smaller than other citrus, have a thin skin, are rounder than lemons, and range in color from yellowish green to green. Lime juice is a large commercial item. The island of Dominica in the West Indies is the major source, and the juice is usually bottled with sugar added to be used in drinks. Limes have the same cooling properties as lemons but their taste is different. Sweet limes are also grown, and Egypt leads the world in both sour and sweet lime production. Key limes are a particular type grown in the Florida keys and along the Gulf Coast. They are mainly used for Key lime pies, a favorite southern Florida dessert.

Note: the lime fruit tree is not the same as the lime or linden tree found in temperate countries. The latter produces no fruit but a flower often used in herbal remedies.

The expression "in the limelight" refers not to the fruit but to the chemical lime through which the light was projected.

LITCHIS (see Miscellaneous Fruits)

LOGANBERRIES (see Blackberries)

MANGOES

The mango is a luscious, rich, spicy fruit that is grown and used widely in the tropics. It is India's best-loved fruit (the country produces 5 million tons a year), and its history goes back six thousand years, interwoven with the history of the Hindu religion. Buddha himself was presented with a mango grove to rest beneath. There is a tale that one time the Buddha was born as a merchant who traded with caravans. He stopped a caravan once at the edge of a forest to warn the traders that poison trees grew in the forest and that they should consult him before tasting any unfamiliar fruit. The traders promised and proceeded into the forest. Within the forest was a village, and nearby a what-fruit tree grew. The what-fruit tree looks exactly like a mango but bears a fruit that is poisonous and causes immediate death. Some of the more greedy members of the caravan hurried to this tree and ate its fruit. The others consulted the merchant, who told them not to touch it. They listened to him, and then he treated the others who had eaten and they recovered. On many other occasions, caravans had stopped by this tree, eaten its fruit, and died. The villagers would then fall upon and loot the caravan. On this particular day, they came expecting the usual spoils and found everyone alive and well. Surprised, they questioned the merchant as to how he knew the tree was not a mango. He said, "When near a village grows a tree / Not hard to climb, 'tis plain to me, / Nor need I further proof to know, / No wholesome fruit thereon can grow!"

The sixteenth-century Mogul emperor Akbar was so taken with the taste and fragrance of mangoes that he ordered an orchard of 100,000 trees to be planted. The trees are also excellent shade trees, often filled with birds (and, unfortunately, mosquitoes). It is possible that the tree is in some way related to poison ivy, and sometimes people sensitive to that plant are also sensitive to the oil in the skin of mangoes.

A mango can weigh as much as four or five pounds. Kidney-shaped fruits are most common, but some are round or long and narrow. The mango has orange skin and flesh when ripe, and a large central seed (stone), and is very sticky. Only ripe mangoes should be eaten and herein lies an irony, for there seems to be more vitamin C in a green mango than in a fully ripe one. There is an art to eating a mango; because the flesh is so sticky and the fibers cling tenaciously to the stone, it cannot be cut in half and eaten like an apple. The result will be too messy and frustrating. Instead, try one of these two methods: cut through the skin and gently peel it back in a broad band. Eat the flesh immediately with a spoon. Or cut the skin in a circle around each end and make several curved lengthwise incisions (around the stone), then peel the skin and lift the flesh in sections. In either event, do not attempt to remove the stone; just eat around it. It's worth the effort!

You can grow a decorative mango tree indoors as long as you don't expect fruit, which takes about four years and probably won't ever appear. Clean the stone from a ripe fruit and plant it "eye" up in soil. Use tepid water (tropical plants are very sensitive to cold) and allow the soil to go dry for a week every month. The plant will probably be dormant for four to six weeks and then a bunch of dark red leaves will appear. You may expect the plant to grow about eight inches in the first year.

Mangosteens   These are fruits, rarely exported, that are related to the mango. They have a tougher rind and a white, soft pulp divided into segments much like a grapefruit.

MEDLARS (see Miscellaneous Fruits)

MELONS

Melons belong to the same horticultural family (*Cucumis*) as cucumbers and squash and, as such, are annual trailing herbs. As we use the fruits, the similarity ends there. They are almost always eaten raw and are probably the most cooling fruit around—so much so that they are often considered a necessity in tropical climates, where, because the water is often considered unsafe for drinking, melon-carriers are a common sight on the streets. It's very simple to transform a melon into a drink: pierce a hole in a partially ripe one and then seal it with a plug of wax (or even the right-size cork). After a few days most of the flesh will have turned to liquid and the melon will be ready to drink. (For something a bit more exhilarating, pour in some wine or vodka before you plug it.)

Although there is some controversy among the nutritional "experts," melons are considered by many to be a good tonic and eliminator. When combined with lemon juice, they are supposed to help in the elimination of uric acid so as to be beneficial in bladder, urethral, and kidney difficulties. (Incidentally, you don't have to have any of these ailments to enjoy a slice of lemon or lime with melon. It's just delicious.) On the other hand, the Greeks and Romans thought melons were beneficial to the stomach and head, and the Romans went so far as to add sharp sauces made from vinegar and pepper to serve with them.

Even the seeds of melons serve a purpose. Orientals eat watermelon seeds like nuts; melon seeds are often dried and strung as necklaces (a great leisure-time activity for children, especially if they paint each seed); and in some parts of the world the eating of seeds is advised to expel tape- and roundworms.

On Mount Carmel there is a field of stones supposedly transformed from melons when a man named Elias ate too many, became ill, and cursed the whole lot.

It is not easy to separate melons into distinct horticultural groups, and for our purpose we will just place the best-known and most popular ones into the general category of muskmelon.

Muskmelons probably originated in Persia, where they received their name because their aroma was so similar to that kind of Persian perfume oil known as musk. The Greeks knew them as early as the third century B.C. and they were brought to Rome about the beginning of the Christian era. They became a favorite of the wealthy (undoubtedly due to the exorbitant price of shipping) and Emperor Tiberius reportedly ate a melon a day.

The Cantaloupe, which was named after a castle in Italy called Cantelupo, was only slightly larger than an apple in Roman times. What we call a cantaloupe in the United States is really a form of muskmelon that originated in Persia. The "true" cantaloupe that grows in Europe has the same pebbled, scaly rind but is not netted like those grown here. Cantaloupes, like most melons, will not sweeten after harvesting but may become softer if kept out of the refrigerator for a few days. Thumping, squeezing, and shaking melons may be an interesting form of excercise but fails as a test for ripeness (See Glossary).

A ripe cantaloupe is delicious, served at room temperature (cold interferes with the delicate taste of most melons), particularly with a spoonful of plain yogurt on top.

Casaba melons are almost totally round, with a rind marked by deep wrinkles. There are two kinds: (1) golden, which has golden flesh and a rind ranging in color from orange-yellow to dark green, and (2) pineapple, with green-white rind and light-yellow flesh.

Crenshaw (also referred to as Cranshaw) melons are round at the base and come to a slight point at the stem end. The gold and green rind is smooth, slightly corrugated, and without netting. The flesh is a bright salmon color and is thick and juicy.

Honeydews have smooth cream-white to grayish skin; the flesh ranges from green to white, and one smell will leave no doubt as to the appropriateness of the name.

Persian melons are usually globe-shaped, with dark-green rind covered over by fine netting. The flesh is orange, very thick and rich-tasting.

Spanish melons are very similar to Crenshaws, with dark-green, slightly corrugated rinds.

Watermelons are classified differently from muskmelons because they are climbers rather than trailers. For years it was believed that they originated in Asia, until Dr. Livingstone found them growing wild in Africa. Watermelons have been cultivated for about four thousand years and were brought to Europe early in the Christian era. European melons are really miniatures compared with those grown in the United States, which often weigh as much as thirty pounds. Aptly named, a watermelon is 92 percent water (and also a lot of seeds, although seedless ones are being grown experimentally). Even though raised in other parts of the country, they are strongly associated with the South. Watermelons are usually symmetrical in appearance, solid green to mottled gray, with bright red flesh. It is common for one side (the side in contact with the soil) to be a bit yellowish. The white inside of the rind is often pickled and used as relish or chutney.

MULBERRIES

Commercially, mulberries are risky items. They must be picked fully ripe; they squash easily and stain readily. The berry has become far more important in song and story. The Psalms said that almighty wrath destroyed the mulberry trees—and this must have been noted as a remarkable instance of divine displeasure, for the mulberry is known not to put forth its buds and leaves until the weather is safe from frost (winter and wrath being thought synonymous). In fact, the mulberry tree has been referred to as the "wisest" of trees and for this reason, no doubt, the Greeks dedicated it to Minerva. Pliny wrote: ". . . when it begins to bud, it dispatches the business in one night, and that with so much force, that their breaking forth may be distinctly heard." Those earlier counterparts of Romeo and Juliet, Pyramus and Thisbe, met secretly under a white mulberry and it was the scene of their tragic end. When Thisbe plunged the sword into her breast, her blood spouted over the mulberries overhead and it was fabled that she said, "You tree, bear witness to the wrongs our parents have done us. Let your berries be stained with our blood in token of their misdoings." Since that time mulberries have been red. (I'm sorry to report, unromantically, that some white ones persist.) Othello's name was Otelli del Moro, from a noble Venetian family originally from Morea, whose symbol was the mulberry (not the strawberry embroidered on that infamous handkerchief). John Milton's mulberry tree at Cambridge supposedly still bears fruit, but Shakespeare's was cut down by the new owner because too many people were annoying him by wanting to see it. An old European superstition was that the devil blacks his boots with the juice of mulberries.

The white mulberry is actually one of the most interesting and important domesticated tree plants. Its importance does not lie in its fruit, which is rather insipid, but its leaves, upon which the silkworm has fed since ancient times. The multicaulis tree (the chief silkworm tree) was cultivated in China, Japan, and southern Asia and reached Europe during the Middle Ages. In the 1820s it was brought to the United States in the hopes of beginning a major silk industry here. Instead, it brought on one of the most dramatic agricultural disasters of American history. Referred to as the "multicaulis craze," of the millions of trees planted hardly any lived, proving too tender for the cold climate. In the East, it still prevails and there is a custom of making a thick preserve of mulberries on the fifteenth day of the first month because a fairy caused an emperor's silk crop to yield a hundred times more in payment for such a delicacy.

In the popular children's song the words were once "Here we go round the bramble bush," instead of the current "mulberry bush."

Medicinally, black mulberry juice is used as a laxative and a refrigerant to cool the blood.

NECTARINES

Despite a common misconception that a nectarine is half peach and half plum, the nectarine is a nectarine—not a cross between anything—but probably, like the peach, evolved from the almond group, which in turn is a member of the rose family. Is that confusing enough?

The nectarine goes back four or five thousand years ago to China and Egypt. Its history is so closely connected with that of the peach that it is easy to understand why many people believe it to be a form of peach. They are very close cousins and have even been reciprocally reproduced. Even though there is little agreement among horticulturists as to the history of the nectarine, Luther Burbank felt that the nectarine was the older form. Whatever their origins, the popularity of nectarines has increased over 400% in the past decade.

Nectarines are still found growing wild in north temperate climates. Their cultivation followed a route from Greece to Rome to Spain to Britain. The modern nectarine in the United States is essentially a new variety, developed in California since World War II.

Nectarines have had their partisans throughout history. In *Cyclopedia of Hardy Fruits* Hedrick described nectarines as being smaller, firmer, and richer in flavor than peaches.

In appearance the nectarine can be distinguished from the peach by its smooth, silky skin which, even after peeling, will still show rose-colored on the flesh underneath. (To peel a nectarine, dip it in boiling water for about a minute, then strip the skin off easily.)

The name derives from the Greek word *nektar,* the drink of the Olympian gods, symbolizing anything delicious and characterizing the fruit perfectly.

OLIVES

> "There is not now a rebel's sword unsheath'd,
> But peace puts forth her olive everywhere."
> Shakespeare, *Henry IV, Part 2*

The olive is the emblem of peace, its branch carried by the dove and linked with the rescue of Noah and his family on Mount Ararat. When

Adam died, three seeds were placed in his mouth and from them grew a cypress, a cedar, and an olive tree. Moses' tears kept the trees alive for the forty years in the wilderness; and eventually they grew into a single tree under which King David wept for his sins. The tree was preserved by Solomon as a relic in his famous Temple. A further story concerning the olive tree is: "Here one day a woman of the Romans, named Maximilia, carelessly leaned against it, then sprang away in fright, crying, 'Jesus Christ, thou son of God, help me!' for flame had leaped from it and ignited her robes. At the call the fire ceased, but the Jews, who had seen and heard, said she was a witch. 'To say that Jehovah had a son is blasphemy,' said they. 'We will hunt this woman from the city.' And they did so. Years afterward the incident came to mind again, for this was the first speaking of the words, 'Jesus Christ.' Finally, the timber was thrown into a marsh, where the queen of Sheba crossed it, to dry ground, when she visited Jerusalem. As her feet rested there a vision arose before her, and she saw Christ suspended on a cross at the hilltop, undergoing shameful death. And so it came to pass, for after a time the log floated to the surface of the morass again, and on the night of the betrayal it was lifted out and shaped into the cross, some say by the hand of Christ Himself" (Charles M. Skinner). Gethsemane, on the Mount of Olives, means literally "olive oil press."

The olive was honored by the Greeks as Athena's tree. She won her contest with Poseidon for the city's patronage by causing an olive tree to rise on the site of the Acropolis. The olive was so revered in Athens that it was used to mark boundaries of estates. In Italy an olive branch is still hung above the door of a new house to keep the devil out. Olive oil was used to light holy lamps in tabernacles ("And thou shalt command the children of Israel, that they bring thee pure oil olive beaten for the light, to cause the lamp to burn always" [Exodus 27:20]) and as a base for expensive perfumes in Rome and Athens.

Olives originated in the Mediterranean region, and France, Italy, and Spain are still major producers; the latter is the largest. Within a twenty-mile radius of Seville, practically all the world's supply of green cured olives is produced. If you should pass through this region, you will notice that after a while the trees lose all definition; the leaves, undercoated as they are with silvery scale, cause the entire countryside to become an olive-green haze. Olive trees require much sun and little rain. They are evergreens that will not bear any fruit for at least eight years (and then only every other year) and that reach their peak of production at thirty-five or forty. They are still young at this point; some have been known to live a thousand years! There are many varieties of olives, differing in size, color, flavor, and oil content. The differences depend upon the locale. Olives are green at first, turning purple and then black as they ripen. They can be eaten in all stages, the green ones pickled in brine. Olives are mostly eaten raw, although many Mediterranean dishes call for cooked olives.

Olive oil is an important by-product of the tree, and some green olives are grown only for their oil, which is pressed from the fruit. World production is estimated at 1 million tons a year. The merit of the oil depends on the quality of the fruit, time of picking, and method of pressing and refining. The finest quality is cold-pressed from ripe fruits, clear, golden yellow, and practically odorless. Inferior grades are more green and are generally used for making lubricants and soaps. If you don't mind getting a bit sticky, one way to judge the quality of olive oil is to put some on the palm of your hand, rub both palms together, then smell the aroma in your cupped hands. The more pungent it is, the less pure it is. "Virgin," when applied to olive oil, means oil from the first pressing. As the pressings and heat increase, the quality decreases. The oil should not be exposed to extremes of light, which will fade its color, or high temperatures, which can make it become rancid. One hundred pounds of fruit will yield between thirteen and fourteen pounds of oil. Olive oil is often used as a base in soaps and shampoos, ointments, and liniments. It is a valuable remedy in bowel ailments, often substituting for castor oil as a mild laxative.

"A long pleasant life depended on two fluids—wine within and oil without."

A Roman saying

ORANGES

"My fruit is better than gold, yea, than fine gold...."
Proverbs 8:19

The name *orange* probably derives from the Old French word for gold, *auranja,* an obvious allusion to the fruit's color and taste. The earliest written mention occurs in the Chinese Book of History (Shih Ching), a collection of documents edited by Confucius around 500 B.C. Certain poets have stated that the golden apples of the Hesperides were not apples at all, but oranges, and that, because Jupiter gave an orange to Juno when they were married, orange blossoms are still worn by brides. Actually, the whiteness and perfume of the flowers would be reason enough.

Oranges probably originated in China and southeast Asia and the rumor of them traveled through Europe thousands of years before the fruit itself arrived. Minstrels chanted of the country in the far-off ocean where "apples of the sun" grew. Before the sixteenth century, many people believed that the orange was a special reward of Mohammedans and should not be eaten by anyone unless they intended to become a member of that faith. This belief seemed logical, since the Arabs introduced oranges into Europe.

When first imported into England, oranges were not eaten raw but were cooked (like most fruits and vegetables at that time) in sauces and relishes. A delicacy was boiled turtle steak served with melted butter, cayenne pepper, and orange juice. The aromatic potential of the orange did not go unnoticed; flowers and rind were used in tea and a particular Elizabethan nobleman had a custom of holding an orange peel to his nose while riding to Westminster Hall, probably because of the stench in the streets. This became a custom, and even today the judges in London perform the rite once a year. In seventeenth-century England the fruit was cultivated in orangeries (usually a greenhouse or other enclosure for cultivating orange trees) and oranges were served raw or boiled with veal, mutton, and other meats. Oliver Cromwell's wife wrote in a book that her husband chastised her for serving his favorite meat dish without its proper sauce (orange). She retorted that oranges were too expensive because of England's quarrels with Spain and that it was a pity her lord did not think of his tastes while engaging in politics.

Columbus brought orange seeds with him to be planted in America, and the huge California orange industry was begun by a Jesuit priest who planted a grove of four hundred trees in the San Gabriel Mission in 1804. Only thirty survived eighty years later, but one lived to a hundred years, as many do in São Miguel in the Azores. Some claim the older the tree, the sweeter the fruit; therefore, the most esteemed are the produce of almost barren trees!

Oranges have always been used therapeutically. Han Yen-Chih, a Chinese writer of the twelfth century, claimed that the peel of the *chu* (orange) was very good when used as a tonic for digestion. In today's pharmacopoeia, orange rind, oil, and flowers are used in a variety of preparations as stomach tonics and sedatives. Orange oil is called neroli oil and, being highly aromatic, is used in perfumes and cooking waters. A Jesuit priest living in Rome in the mid-seventeenth century wrote that a fermentation of orange flowers could be used as a remedy for the heart and that sneezing could be provoked by a snuff made from the rind. As an antiseptic, oranges stuffed with cloves were carried to ward off the plague. I've heard that Arabs restore color to gray hair by drying half an orange and soaking it in oil for a month, but I can't figure out whether they eat it or place it on their heads.

Every part of the orange is used. Margarine and other cooking fats are derived from the seeds, as well as a dye to fix the colors in some synthetic fabrics; the peel is used for terpenes for paint; pectin from the pulp is used as a jelling agent for preserves and also in the treatment of wounds. And when it seems as if there is nothing left to extract, the residue is fed to cattle, and from the water in which the peel is washed comes tons of molasses.

In many countries the inner white peel (which has some laxative action) is kept intact so that both fruit and peel are eaten together. An orange aids in the digestion of milk when combined with it (see *Lemons*) and a raw egg yolk whisked in greatly increases the nutritional benefits. Some nutritionists say that whenever possible oranges should be eaten alone—not in combination with other fruits; although juices may be combined.

". . . before ever we start our observations, we think we know what it is the duty of reality to be like. For example, it is obviously the duty of all oranges to be orange, but, if in fact they aren't orange, but, like the fruits of Trinidad, bright green, then we shall refuse even to taste these abnormal and immoral caricatures of oranges."

Aldous Huxley

Nature and man are often at odds and so, despite the name, many oranges are ripe when green. (All oranges are tree-ripened; they will not ripen after being picked, like some other fruits.) A drop in temperature can cause an orange to turn from green to orange. To come up to their "image," green or yellowish green oranges are often placed in a "degreening" room where ethylene is used to decompose the chlorophyll (green), giving the fruit an orange pigment. Unfortunately, the process sometimes goes a bit further. The oranges are scrubbed and much moisture is removed (so that they will keep better) and if they're still not orange enough, they are sometimes dyed and waxed—all for appearances! If you see a greenish orange or one with green spots, don't hesitate to buy it—it's probably more natural.

Oranges fall into two major categories: sweet, and bitter or sour. Within these, there are many varieties. Sweet oranges are more popular for out-of-hand eating and juicing. Some of the best known are Jaffa (from Israel), Valencia, and blood oranges (with a reddish pulp) such as Maltese. Sour oranges (Seville) are used mostly for making marmalade. The Bergamot orange is grown in the Mediterranean region almost exclusively for the extraction for perfume and soap oil. Generally speaking, heavy, thin-skinned oranges contain more juice; thick-skinned ones are easier to peel and section. The navel orange, originally developed in Brazil, can be distinguished by a navel-like bump at its apex, is practically seedless, and is very popular in the United States. Southern California and Florida produce most of the oranges sold in the United States because the weather there is warm and dependable enough to eliminate the danger of frost, which kills most citrus. In the world market the other major producers are Spain, Israel, South Africa, Brazil, and North Africa.

PAPAYAS

The papaya is a native American. This tree melon (actually the tree is really a giant herbaceous plant) has an elongated shape, yellow or orange skin, luscious, glossy orange flesh, and many black seeds, which look like oversized caviar. The fruit can range from one to twenty pounds and will mature in eighteen months from the time the seed is planted. Its easy propagation by seed caused its rapid spread to other tropical countries from its native West Indies. In the United States, Florida is the largest producer. Although only occasionally seen in markets here, in other countries the papaya ranks as one of the most important tropical fruits. It is also known as pawpaw.

The papaya plant contains an enzyme called papain, which breaks down protein in much the same way as does the animal enzyme pepsin. This enzyme is found in the fruit, stem, and leaves and is extracted by cutting the surface and collecting the white latex fluid underneath. This fluid is dried into powder and used in chewing gum, toothpaste, and sometimes medicinally as a digestant. (Papaya juice is also particularly effective as a digestant.) Dried papain is often the base of meat tenderizers, and a tough steak wrapped in a bruised leaf and then cooked will become amazingly tender. Papaya seeds are sometimes used as a spice or chewed. The raw fruit is a delicious and cooling treat.

PEACHES

> "The ripest peach is highest on the tree."
> James Whitcomb Riley

Peaches originated in China, where they first grew wild. Cultivation began

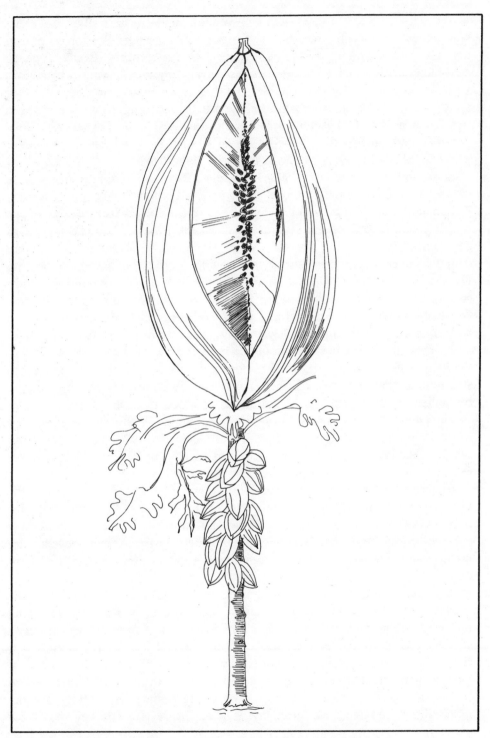
The juice made from papaya is delicious and an aid in digestion.

in the tenth century B.C. Peaches moved along caravan routes to Persia, where they received their name (Latin *persica,* meaning "Persian"), and at one time were known as Persian apples. In first-century Rome a peach cost the equivalent of $4.50, and in Victorian England it was still precious ($5.00) and was very fashionable to serve at a dinner party in a nest of snowy cotton. The Spaniards are believed to have brought peach seeds and trees to America. Columbus carried them on his ship, and wherever Europeans settled, trees were planted: by the French in Louisiana; by the Pilgrims in Massachusetts; by the English in Jamestown.

Some modern varieties of peaches are not so very different from those of ancient cultivation, but those eaten in the United States have all been developed here. Commercial growing began in the early 1800s, and Georgia, which produces one of the famous peaches, the Elberta, became known as the Peach State. Peach growing is so popular, however, that trees are found in every state, with twenty-six having more than a million bearing trees each. In fact, the United States produces one-half of the entire world's supply and there are thousands of named varieties in this country alone. Most commercial packers "defuzz" peaches during packing, and some even wax them after that to prevent water loss and extend shelf life.

Peaches grow widely throughout the temperate zone and are usually self-fertile, the flowers on a tree pollinating themselves. The tree is so close to the almond that the two hybridize easily and almond kernels impregnated by pollen from a peach tree can often produce peaches. There are two broad types of peaches, clingstone and freestone. Within these categories there is a great variety in skin and flesh color, which may be greenish white or yellow. White peaches are considered the hardiest and best for cooler climates; yellow are the most popular.

In plant medicine, the oil from the stones (persic oil) is used in the manufacture of fine soaps and other toilet preparations as a substitute for almond oil. From the bark and leaves an infusion is made for treating gastritis, coughs, and chronic bronchitis. Taken by themselves, peaches are an excellent eliminator and contain fewer calories than either apples or pears.

A popular folktale of Japan goes as follows: an old woman, while washing her clothes at the river, saw a large round pink object splashing and rolling about in the water. She fished it out and found it was a peach so large that it would feed her husband and herself for several days. When they broke it open, inside the stone they found a tiny child. They brought him up with great love and gave him everything they could afford. When he was grown, he invaded the Island of Devils, where he defeated the demons and seized their treasure, laying it at the feet of his beloved foster parents. The Chinese goddess Si-Wan-Mu has a tree that is the center of the Taoist paradise (Confucius called the peach *tao*). Once every three thousand years its fruit ripens and confers the gift of immortality on one who eats it. A peach tree was called the tree of life by ancient Chinese writers, and because it often symbolized longevity, it was depicted on plates and bowls used for birthday celebrations. Pink peach blossoms somehow connoted promiscuity to the ancient Chinese, and gardeners were warned against planting peach trees too near a lady's bedroom window. In American slang "peach" has come into use as a complimentary name for a good-looking girl; the word "peachy," although old-fashioned, still means something good; and a "peaches-and-cream complexion" is just that.

PEARS

Pears originated in central Asia centuries before Christ, but it was Homer who brought them to the world's attention. In the *Odyssey,* he heralded them as "a gift of the gods." Marco Polo remarked that the Chinese had carried the cultivation of pears to perfection and produced fruits that were white on the inside, which melted upon eating, and weighed ten pounds each. The Romans brought the pear to England, where the monks grew many varieties. There is a story that King John was poisoned by a dish of pears, offered as a delicacy by the monks of Swinsted. Cooked pears (particularly poached) were very popular in France during the middle of the nineteenth century as an elegant dessert item, and this tradition still remains strong in French cuisine.

Pears, like many other fruits, were introduced into California by Franciscan monks who planted them in mission gardens there. Perhaps the most extravagant pears in history are those grown on the island of Jersey, which can bring a price of seventy dollars or more for a dozen in the London market. The reason: each tree bears only a few precious fruits. After three hundred years, trees in Herefordshire, England, are still bearing fruit.

There are more than three thousand varieties of pears, but only about twenty are popularly sold. These can be divided into two groups: European, which are soft and include the well-known Bartlett and Anjou; and Chinese or Asian, which are hard, slightly gritty, and sometimes called sand pears. The latter are preferred for preserves. Many hybrids have been developed, notably the Kieffer, Le Conte, and Seckel. Pears come in a variety of colors: green, yellow, brown, and all manner of combinations and sizes. In Europe they are grown mostly in France, Belgium, and Germany;

in the United States they are cultivated in New York, throughout New England, and on the West Coast. In earlier times, pears were budded on pear stock and took a long time to crop. Today, most are grown on quince rootstock.

Perry is a drink (produced mainly in Normandy) equivalent to apple cider and is made from the fermentation of wild pears or their cultivated counterparts. Perry was drunk by the Romans after eating mushrooms as an antidote to any possible poison contained in them and is still a good accompaniment to that vegetable. There is some evidence that it is a good drink for diabetics because pears contain less sugar than most other fruit. There is a tale that water from a well sunk close to a wild pear tree can cure gout.

In addition to *perry*, some other words in English pertaining to pear are *pyrus* and *pyriform,* the latter meaning "pear-shaped"—perhaps a genteel way of referring to women with that characteristic shape.

PERSIMMONS

". . . The fruit is like a medlar; it is first green then yellow and red when it is ripe: if it is not ripe it will drive a man's mouth awrie with much torment; but when it is ripe, it is as delicious as an apricock."

Capt. John Smith, Jamestown

There are two species of persimmon, American and Japanese (the latter is called kaki). Each has an irritating astringency unless fully ripe. Persimmons contain an acid called tannin, which causes this sourness, but the Japanese combat it in the following manner: the green fruit is placed in tubs that once held sake (rice wine) and tightly covered. The presence of the alcohol was considered helpful in removing the astringency, but time proved that just sealing the fruit tightly was enough. Persimmons are also rich in organized pepsin and for this reason are excellent for the digestion.

Persimmons require a warm temperate climate and grow extensively in their native China and Japan. Both the oriental and American kind are grown here, mostly in the southeastern states; the American species sometimes turns to a dark red-maroon. There is a small commercial crop in California, but often persimmons are grown in home gardens and many are picked from wild trees and not cultivated at all. I have heard that when the seed of a wild persimmon is carefully opened it will be seen to contain a tiny, perfect knife, fork, and spoon set!

The pineapple is formed from hundreds of small blossoms.

PINEAPPLES

Technically, the pineapple is not a fruit, but a cluster of berries, formed from hundreds of individual blossoms. This succulent plant is native to South America; but the most important pineapple-producing region is Hawaii. In early colonial days, it was a common practice to carve and paint stylized pineapples on furniture, partly because the fruit is so decorative and also because it was a symbol of hospitality. Though this custom has disappeared, a pineapple still makes a festive centerpiece for the table.

The name *pineapple* comes from the obvious resemblance of the fruit to a large pinecone. The Carib and French name is *ananas,* which, despite the sound, has nothing to do with bananas but comes from *a,* meaning "fruit," and *nana,* meaning "excellent."

Nutritionally, pineapples contain a protein-digesting enzyme similar to that in papayas. They have great value as a digestive aid, and for this reason are an excellent first course of a meal. Sprinkling juice on meat before cooking will serve to tenderize it, and the custom of cooking ham with pineapple slices was probably as much practical as attractive. The only problem is that when pineapple is cooked, it loses some of its digestive properties, so eat it raw, too.

Pineapples reduce heat and are excellent in treating fevers. The Indians fermented them and made a liquor. Externally, they can be used to dissolve corns and treat skin ailments. The tough spiny skin of the pineapple prevents insecticides from getting inside, so that all pineapples, whether organically grown or not, are relatively free from this danger. The strong, flexible fiber from the leaves is used in the manufacture of a regional textile—the delicate, soft, transparent piña cloth, which is often used for scarves.

Pineapples must have sandy soil and temperatures of seventy to eighty degrees to grow. Plants do not grow from seed, but from slips or by planting the entire crown of leaves. You can experiment with a house plant by cutting off the crown and about one inch of the fruit, drying it thoroughly for a few days, and then planting it in a shallow pot (because its roots are shallow). The potting mixture should be one-half soil, one-quarter humus, and one-quarter dried coffee grounds.

PLANTAINS

Plantains closely resemble bananas, but actually they are eaten more like a cooked vegetable than a fruit. Plantains are longer and thicker than bananas, less sweet and more starchy. They are often sold green, blemished or spotted, and are never eaten raw. They are a staple in many tropical diets, particularly in East Africa, and are a favorite particularly with Puerto Ricans and Cubans, who call them *plátanos,* and cook or roast them. The American Indians used the plaintain to cure poison ivy.

PLUMS

The founder of Taoism, the oriental philosophy that urges man to follow nature and not interfere with the natural goodness of the human heart, was Lao-tse, supposedly born white-haired under a plum tree in 604 B.C. Although many other trees form a rich orchard in Chinese mythology, the plum is singularly associated with great age (and, therefore, wisdom). Lao-tse's family name (Li) means "plum tree." Plum blossoms carved on jade symbolize the resurrection. It is a great irony that this tree, the fruit of which was so symbolic of China's Golden Age, became an important food for the invaders, the Huns and the Mongols, who brought such ruin to that country. Plums have been cultivated for more than two thousand years, but wild ones grew in the northern hemisphere from the West Coast of the United States to Japan. When the Asiatics moved into Europe, they brought with them dried plums (prunes) and also introduced plum trees.

Two general types of plum grow in the United States: Japanese and European. Cultivation of the European plum began in the United States around the beginning of the seventeenth century, of the Japanese around 1870. Plums are blue, red, yellow, green, and combinations thereof. They may be round or oval. There are many species with varying uses (see Glossary). Luther Burbank alone developed sixty varieties of plums. The greengage was named after Sir William Gage, who brought them from France to England; the damson (used mostly in preserves) is named after Damascus; and there is a plum named tragedy, which I think has much to overcome. One type of plum is used only as a flavoring for sloe gin, which is flavored with plum rather than the standard juniper berries. A kind of cherry plum, whose flowers closely resemble peach and apricot blossoms, is used ornamentally with its leaves and flowers.

The background story behind Little Jack Horner's Christmas pie is most revealing. A legend claims that Jack Horner was a certain Thomas

Horner who was a steward to the last of the abbots of Glastonbury Abbey. When Henry VIII began to take over all properties belonging to the Church, the abbot dispatched his steward to the king with a Christmas gift: the pie, containing the deeds to several estates. On the way, sticky-fingered Thomas supposedly opened the pie and pulled out a deed to one of the manors—and therefore a "plum," giving us the meaning of something choice that still lingers in our vocabularies today.

Prunes All prunes are plums, but not all plums are prunes. A specific type of plum tree produces fruit that can be dried without fermenting and still contains pits. An area around the Caucasus Mountains, near the Caspian Sea, was probably the place of origin of these plum trees. The Huns, Turks, Mongols, and Tartars used prunes as a staple in their diets, probably because the fruit traveled so well in its dried state and was extremely nutritious. The best known prune is the D'Agen, known as the French prune, which has a small pit and is grown mostly in California. That state provides 98 percent of the United States' supply of prunes and 69 percent of the world's.

Prune plums are deep blue to almost black when ripe. The process of drying is as follows: they are dipped in hot liquid to crack their skins and then placed in cold water. Some are dried in the sun for four or five days but most, like other dried fruits, are dried in dehydrators for about twenty-four hours and the process is often speeded up with the addition of sulfur (try to buy naturally processed prunes). When they are cooked, prunes are restored to their original state; the dry weight is nearly doubled by the absorption of water. Prunes may be soaked before cooking, but it is not necessary to do so. Simmer, do not boil them; keep the lid on and don't stir, to avoid breaking the skins. Allowing them to stand in cooking liquid (with a touch of claret or sauterne added a few minutes before the finish) makes them softer and plumper. The cooking water should be used, and, of course, prune juice is a nutritious and delicious beverage. Prunes have a reputation as a natural laxative, probably because their cellulose content is very high and foods high in cellulose tend to regulate the bowels. Prunes also assist in the elimination of mucus. It should be noted that prunes (as well as plums) produce an acid reaction in the body, rather than the usual alkaline one produced by most other fruits.

POMEGRANATES

The pomegranate evokes fragrant remembrances of mythical and biblical stories, for this fruit, perhaps more than any other, has symbolized the exotic or rare. It could have been either the tree of life or the tree of knowledge. Certainly it is as old as the stories of the Garden of Eden. King Solomon, who had it carved on the brass pillars of his temple, sang of an "orchard of pomegranates." Moses described the promised land as "a land of wheat and barley, and vines and fig-trees, and pomegranates." A favorite in legends and fairy tales, the pomegranate is the mystic fruit of the Eleusian rites, called the fruit of hell. In one Greek myth, young Persephone (Proserpina) was gathering flowers when the earth opened up and Dis or Pluto, god of hell, carried her off to be his wife and queen in the gloomy subterranean world. Demeter (Ceres), the goddess of the earth and mother of Persephone, left Olympus and went to live in Eleusis, where, in her anger, she would allow no seed to grow on earth until her daughter was returned. Zeus ordered Pluto to bring back Persephone. He obeyed but, just as she was leaving, begged her to eat the pomegranate he gave her, and by so doing she was forced to spend one third of the year—the winter months—in the dark netherworld. This legend has been interpreted in many ways: the pomegranate symbolizes the power of the world of darkness, the fruit of which, once experienced, deprives man temporarily of the light and warmth of immortality. The pomegranate also becomes an archetype of all fruits that germinate beneath the earth and bloom in the spring. (Paradoxically, pomegranates grow on trees.)

In China, the pomegranate's significance is chiefly as a fertility symbol. Women desiring children offer the fruit to the goddess of mercy, and porcelains in her temples are decorated with pictures of pomegranates. In Turkey a bride throws the fruit to the earth to smash it. The number of seeds expelled is supposed to indicate the number of children she will have. Bacchus turned the maid whom he seduced into a pomegranate tree. On its fruit he placed a crown to compensate for the one which he had promised her. The tree is very distinctive, with its beautiful, vivid orange-red flowers and glossy leaves, and was undoubtedly Shakespeare's inspiration when he had Juliet say, "It was the nightingale, and not the lark, / That pierced the fearful hollow of thine ear; / Nightly she sings on yon pomegranate tree." Travelers from the East had reported seeing choirs of nightingales in pomegranate trees. Granada in Spain, which was probably named for the fruit, had the Avenue of Pomegranates planted by the Moors, who made a split fruit the coat of arms for that city.

The word *pomegranate* means "grained apple," which becomes obvious upon opening one. If not approached with the greatest care, a pomegranate is not an out-of-hand fruit, but more an on-the-shirt-face-and-walls one. My strong advice is: don't cut it in half. I've heard various solutions,

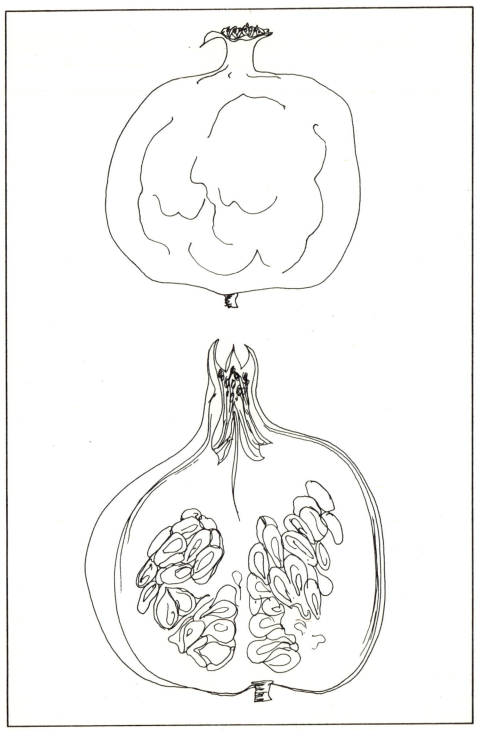
The red juice is drawn from the seeds of a pomegranate, but the white inner core is usually not eaten.

some of them very fanciful; here are a few for your experimentation. (1) Hold the fruit with the stem end toward you and chop off a deep slice from the other end—the one that once wore Bacchus' crown. You should then be able to peel back the outer skin and divide the fruit into its natural segments. Most people suck the seeds and eat only the red, juicy seed coverings, called arils, and leave the white core. (2) Squeeze the fruit in your hand until it gets somewhat soft. Pierce a hole at one end and drink the juice. If you're disinclined to risk any technique, you can buy pomegranate juice to drink "straight" or use in meat dishes, stuffings, and curries. Pomegranates are also made into grenadine syrup, and the wine of which Solomon spoke is still produced in Iran today. The first known sherbet was a preparation of pomegranate juice and snow.

In addition to its use as a refreshing fruit, pomegranates have been used medicinally throughout time. The seeds were used by Pliny as early as A.D. 50 to expel tapeworm, and one of the oldest drugs, known as taenifuge, is made from the bark of the tree and is still used for that purpose. The rind is valuable in treating chronic dysentery, but large doses may cause vomiting, cramps, and a discouraging list of side effects. Some herbalists claim good results in strengthening gums, loose teeth, and ulcers in the mouth and throat, and in curing anemia, all of which seems almost to classify the pomegranate as a panacea. In addition to its medical value, the rinds of the fruit have been employed in the tanning of leather. It is not surprising when one sees the abundance of pomegranate trees growing in Spain that the Moors began tanning fine leather (cordovan) with it in Cordova. Even the flowers are utilized to make a strong red dye. The seeds (or more explicitly, the white core that you don't eat) can be sprouted and planted to make a lovely house shrub, which may even flower given a lot of warmth. (Pomegranates grow in tropical and subtropical areas like Saudi Arabia, Israel, South America, and certain Mediterranean countries.)

For the astrological eater, as pomegranates are an alkaline agent, they are supposed to be particularly helpful to Aquarians. I wonder if that makes them the fruits of the Aquarian Age. Perhaps the pomegranate's greatest honor was bestowed when Mohammed said, "Eat the pomegranate, for it purges the system of envy and hatred."

QUINCES

Quinces look a little like apples and pears and are a fairly neglected fruit now, although they are used as a rootstock for pears. Native to western Asia, Persia, and Greece, they have been cultivated since ancient times, when they were evidently looked on with much more respect, some even claiming them as Venus' golden apple. (But, in this regard, they joined company with several other fruits.) Nonetheless, quinces, as tokens of love and happiness, were eaten or exchanged by lovers, and to dream of quinces was interpreted as a sign of success in love. Pliny the Elder wrote of the belief that a quince warded off the evil eye, and in medieval times they were served at wedding feasts. The name *quince* in English came about as a corruption of an ancient city of Crete named Cydonia where the trees grew wild. In Portuguese the name is *marmelo,* and one of the more common uses of the quince is in marmalade. Being hard and acid, it is usually cooked with sugar and sometimes into a candy called contignac.

"It hath been found most true that the very smell of a quince hath taken away all the strength of the poison of white hellebore."

Nicholas Culpeper

White hellebore seems fairly rare, but diarrhea and dysentery are not, and quince seeds can come in handy. For eye diseases, a soothing lotion can be made from the quince for external poultices.

Probably because of its taste, the poor quince has come into slang usage as a term for a particularly unpleasant, sour person—usually a woman. No comment!

RASPBERRIES

The botanical name for the raspberry means the "bramble of Ida," probably because it was said to have grown in profusion on Mount Ida, where Paris awarded the apple to Venus. There are many varieties of raspberry, native to most temperate zones of the world. They can be red, white, yellow, black, or purple and are recognized by the rough and bristly appearance of the fruit. Many species have remained uncultivated, because raspberry seeds are widely distributed by birds. For hundreds of years, herbalists have recommended raspberry-leaf tea for pregnant women to help avert miscarriage and to help ease their labor pains, increase their milk supply, and generally give them strength. Raspberry vinegar was used for sore throats, and out of the leaves one can make a poultice for cleansing wounds. The American Indians crushed the roots of the black raspberry plant and boiled

Quince bears a strong resemblance to a pear, but is not popular raw because of its acid taste.

them for treatment of dysentery. One current herbal-medicine writer claims that even eating the berries "straight" will help in a bout of diarrhea. A richly flavored honey is derived from the flowers.

STRAWBERRIES

"God could doubtless have made a better berry, but doubtless he never did."
Jonathan Swift

The strawberry has been called "the oyster of summer," its delicate refreshing flavor synonymous with those golden days. Wild strawberries probably grew in temperate countries since ancient times; an early American colonist wrote, "We cannot sett downe a foot but tred on strawberries," and they are still found frequently in woods and shaded areas. If you've ever been fortunate enough to eat *fraises de bois* in France, no description of their taste is adequate. The larger modern cultivated strawberry was developed in the United States from a Chilean species by way of France. This was in the beginning of the eighteenth century, but it was not until the middle of the next century that commercial production was made possible by developing a firm-fleshed fruit that could be shipped without damage.

> "The strawberry grows underneath the Nettle,
> And wholesome berries thrive and ripen best
> Neighbor'd by fruit of lesser quality."
> Shakespeare

Strawberries are really perennial herbs, the fruit of which is an enlarged receptacle on whose surface the seeds are imbedded. There seems to be much difference of opinion concerning the derivation of the name. Various theories are: (1) it came from the practice of laying straw under plants to prevent ripe berries from being soiled by mold; (2) the specks covering the surface of berries resemble straw; (3) the Anglo-Saxon word for *stray* was *strew,* an apt description because while the strawberry's roots stay in one place, its shoots run off in all directions.

The strawberry is dedicated to the Virgin Mary. There was a superstition that if a mother came to heaven's gates with the stain of strawberries on her lips, she was cast down to hell because she had trespassed in the Virgin's fields. At one time some believed that infants ascended to heaven disguised as strawberries and people on earth avoided the fruits lest they be considered cannibals. (Less superstitiously and more practically, they could well have been allergic; there is something about strawberries that induces hives in many people, especially when the berries are combined with citrus fruits.)

Among herbalists, strawberry leaves are used in a tea for the treatment of diarrhea and some urinary infections, also to relieve aches in the hips and thighs. Strawberries are held to have an excellent effect on the teeth and gums; they do everything from whitening to lightening. If you're a Libra, they're supposed to be particularly helpful for maintaining equilibrium. Used in cosmetic lotions and creams, strawberries help to remove superfluous skin, bleach out freckles, and firm the skin; and even if you don't believe this, the smell is delicious.

Strawberries are easy plants for the home gardener and will grow in a variety of inventive receptacles, such as large barrels. They are very popular: there are at least two varieties of strawberries grown in the United States familiarly named Joe and Jessie.

How is this for a unique recipe? In Lapland they make a special Christmas pudding called *kappaltialmas,* which consists of strawberries mixed with reindeer milk and then dried in the form of a sausage.

TANGERINES

Tangerines are known by many names in other parts of the world. Some of them are king orange, satsuma, Naartje, and mandarin, the last denoting its origin in the Far East. Tangerines were introduced into this country late, around the middle of the nineteenth century, but have become very popular as an out-of-hand fruit because of the ease with which the loose rind ("zipper skin") peels and the segments separate. Tangerines are usually smaller than oranges and of a deeper orange or reddish color. The name supposedly comes from the city of Tangier in Morocco. At least one girl, born in Tin Pan Alley, has been named after the fruit "Tangerine, she is all they claim . . ." (but probably only because it rhymes with Argentine).

UGLIS (see Miscellaneous Citrus Fruits)

WILD BERRIES

Described below are some of the more common edible wild berries.

Elderberries are probably best known for the dark purplish red wine, good for colds and other things (remember *Arsenic and Old Lace*?). Less common uses involve elder flowers, from which a wine is made, and which can be dipped in batter and fried (see *Squash*). The elder tree is considered a great healer, and a famous physician was reported to have held it in such reverence that he took off his hat each time he passed it.

Barberries have been used since Christ's time as an ornamental hedge. They have never been too popular as a fresh berry but are cooked in sauces and pies. In northern Europe barberries were used sometimes as a substitute for lemons in drinks. Medicinally they have many uses. The famous herbalist Nicholas Culpeper wrote: "Mars owns the shrub and presents it to the use of my countrymen to purge their bodies of choler."

Juniper Berries

> "Lay there by the juniper
> While the moon is bright,
> Watch the jugs a-fillin'
> In the pale moonlight."
> © Albert F. Beddoe, "Copper Kettle"

Juniper berries are best known for the flavor they impart to alcohol, particularly gin, but even the teetotaler has been known to benefit from them, because they have a long history as an herbal diuretic since ancient Greece.

MISCELLANEOUS FRUITS

Medlars, like quinces, are in the rose family. Medlars often grow wild in Europe and are a very strange-looking fruit whose five seeds show through in the receptaclelike tip. The fruit is brown and is eaten raw or made into jam.

Dog Rose is more commonly known as rose hips. It is the fruit of the rose after the flower has faded and the petals fallen, and is bright red when ripe. Rose hips are valued not so much as a fruit but for the syrup and tea made from them. The ancient Greeks referred to them as the food of the gods, and it was discovered about twenty years ago that they contain a higher amount of vitamin C than any known food.

Acerola Berries The tree on which acerola berries grow is known in the tropics as the health tree. The fruit is small and cherrylike and was often used as an ornament. Its flavor is similar to an apple. Like rose hips, acerola berries contain a very large amount of vitamin C and are mostly used in concentrated juices and powders. They are often combined with rose hips as ingredients in natural vitamin C tablets.

Litchis *Litchi* is sometimes spelled *lichee* or *lychee*. Litchis are often called lichi nuts because a favored use is drying the fruit, which then becomes shrunken, brown or blackish, and nutty in taste. The outside becomes a brittle, thin dark shell. When fresh, it has a bright scarlet warty skin. The fruit is native to China and, although also grown in Florida, it is associated for all time with Chinese cuisine as that white, jellylike, acid-sweet small fruit you order for dessert. There is a story that, while living in exile in Canton, a famous Chinese poet declared, "Litchi would reconcile one to eternal banishment."

Loquats Despite the similarity in name, loquats are not related to kumquats but are really medlars, known often as the Japanese medlar. The fruit grows in China, Japan, and north India, but has been cultivated in the Mediterranean region as well. Its color ranges from yellow to deep orange. It is pear-shaped and the size of a crab apple, with a sweetish acid flavor. It is very juicy when ripe and is eaten fresh, stewed, or made into preserves. In Bermuda, a liqueur is made from loquats.

MISCELLANEOUS CITRUS FRUITS

Citrons The citron was probably the first citrus to reach the Mediterranean region from the Far East, around 300 B.C. It is still grown mainly in Greece, Sicily, and Corsica. The citron is more elongated than circular, and the greenish yellow skin is often rough and warty. The part of the fruit almost exclusively used is the thick white inner skin, which is made into candied peel. A special type of citron called *etrog* has long been used in accordance with the Mosaic law of the Jewish religion in the Feast of Tabernacles, when it is passed around to each member of the congregation who smells it and then praises God for the sweet odors He has given man.

Kumquats The kumquat is technically not a true citrus, but is very closely related. The fruits look like tiny oranges. The rind is spicy and sweet, the flesh is tart, and the entire fruit may be eaten. Kumquats are often used to make marmalades and jellies, and are also candied. The name comes from a Cantonese pronunciation of the Chinese word *chin-chu,* which means "golden orange," and the fruit is primarily produced in China, where its resistance to cold permits it to grow farther north than any other citrus. At elegant banquets in the Orient, the guests receive little dwarfed trees and pluck their dessert from them. In the United States, in more prosaic fashion, the kumquat is often served with toothpicks at the close of a Chinese meal.

Limequats, Orangequats, and Citrangequats are hybrids of kumquats with other citrus. The main purpose for the cross-breeding is to produce fruit with greater resistance to cold.

Uglis The ugli is a hybrid grapefruit-tangerine. The rind resembles a small, rough-skinned grapefruit and the inside segments look like those of the tangerine. It is a fairly uncommon fruit. Another cross between these two fruits is the Tangelo, which resembles an orange more than a grapefruit and is a bit more attractive than the ugli (the pun is intentional).

Pomelos The pomelo (pummelo) literally means "melon-apple," because it looks like a tree melon and is the ancestor of the grapefruit. It is the largest of the citrus fruits (fifteen-pound ones were used for decorating banquet tables) and is supposed to have originated in Malaysia over two thousand years ago. The fruit is sometimes called a shaddock after the ship's captain who supposedly introduced it into the West Indies. Pomelos have thick skin and fairly bitter pulp and are not often found outside of those areas in southeast Asia and the tropics where they grow.

Tangors are hybrids between tangerines and sweet oranges. The popular Temple orange is a tangor.

Clementines The clementine is considered a tangerine by some and a cross between a tangerine and a sweet orange by others. In any event, it is the size of an orange, with a peel easily removed like a tangerine's, and is grown mostly in North Africa.

MISCELLANEOUS TROPICAL FRUITS

Breadfruit In this part of the world our only association with breadfruit may

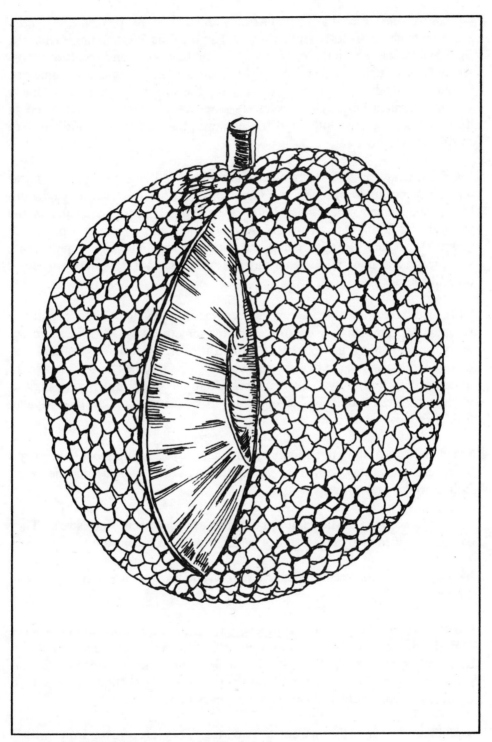

When roasted, breadfruit tastes a bit like bread.

be that we remember that Captain Bligh's mission in *Mutiny on the Bounty* was to introduce breadfruit plants from Tahiti to the West Indies. Generally the breadfruit is not found outside of the Pacific islands and southeast Asia, where it is a staple food for the eight months it is in season. It comes by its name naturally. This huge fruit (usually the size of a man's head) has a starchy, seedless pulp, which is rarely eaten raw. Instead, it is roasted in its skin between hot stones until it achieves the consistency and taste of freshly baked bread.

Cherimoyas are part of the family commonly called custard apples, a name derived from the consistency and flavor of the fruit. The cherimoya resembles a small green pineapple, without the crown. It is found locally in the Central and South American and Asian tropics and is rarely exported to temperate regions. Besides being a delicious combination of a pineapple, strawberry, and banana, the cherimoya is often used for dysentery and other digestive ailments.

Kiwi Fruit is a tropical fruit that has become more available and popular in recent years. This small fuzzy brown fruit is sometimes called Chinese gooseberry, but it is not a gooseberry at all and it comes from New Zealand (although it did originate in the Yangtze valley). The luscious flesh is chartreuse in color, with tiny brown seeds running along the sides, and is very high in vitamin C. Kiwi is delicious eaten plain, in salads or pies, or, like papaya, used as a meat tenderizer.

Sweet Sops are in the same family as the cherimoya and are more "custardy." They are sometimes called sugar apples and are popular in the West Indies, where they grow.

Sour Sops are also in the same family, but are more acid than sweet. They can be identified by their soft spines, which grow on a green rind.

Ilamas   The ilama resembles the cherimoya and grows in Mexico and Central America.

Passion Fruits   The passion fruit, also known as purple granadilla, grows in some Mediterranean countries as well as in the tropics and can sometimes be found in markets in Europe. If you find one, taste it. Despite its outside appearance, which is usually shriveled and unappealing, the inside is a juicy, sensuous pulp that one can eat, seeds and all.

Sapodillas grow mostly in Central America. They are brown fruits with a

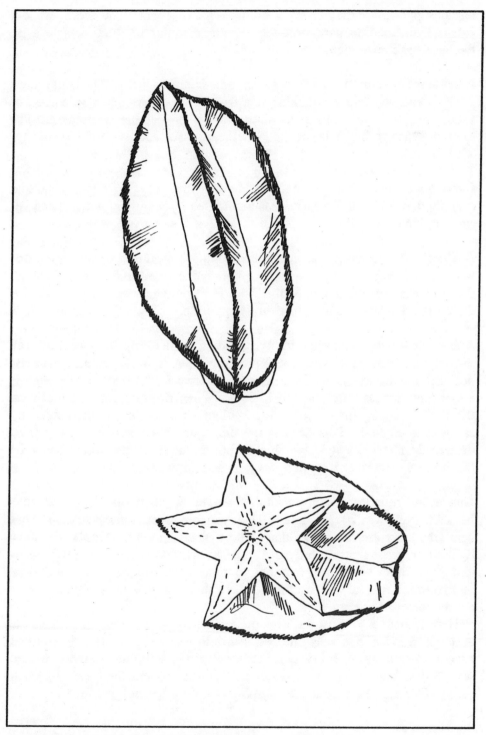

Carambola is known, colloquially, as star fruit.

luscious pale pulp that tastes a bit like brown sugar. The seeds are not eaten. The sapodilla bark produces a more familiar product—chicle gum, the basis for chewing gum.

Rambutans Rambutans belong to the same family as litchis. The fruit is most unusual and one might not think it a fruit at all but rather a spiny red or yellow flower. Inside this plum-sized fruit is an eatable white flesh. The rambutan grows in Malaysia and southeast Asia and is rarely seen anywhere else.

Carambolas are yellow, oblong, ribbed fruits that grow in Indonesia. Occasionally they are sold in other areas and are easily identified when sliced because inside is a beautiful five-pointed star of acid-sweet yellow pulp.

Durians The durian is a very peculiar-looking fruit, full of spines. The inside is a creamy pulp that has an almost sewagelike smell and taste. Although the southeast Asians eat it abundantly, few foreigners acquire a taste for the durian. Understandably, it is rarely exported.

Akees If you've ever visited the West Indies, particularly Jamaica, and seen this popular fruit there, you would probably be able to describe it perfectly, because it is clearly one of the most distinctive fruits grown. The akee is about three inches long and red in color. When ripe, it opens naturally on the tree, revealing three round, black olivelike seeds, each surrounded by cream-colored flesh. This flesh is the only part eaten and is usually fried or boiled. The red exterior is highly poisonous, and great care must be taken to eat only ripe fruits that have opened naturally.

Guavas are round or oblong, usually pale yellow when ripe. The flesh may be white, yellow, or salmon red. They can be sweet, mildly acid, or very acid (the latter are sometimes made into a vinegar used by people who have difficulty digesting other vinegars). The guava contains many small seeds and is very juicy when ripe. It grows abundantly in Mexico, the West Indies, and Hawaii but is rarely found in northern markets. It is most often eaten as an ingredient in preserves.

Here is a story to illustrate how pungent the odor of ripe guavas can be. A large wooden box was shipped from Florida to New Jersey marked GUAVAS. PERISHABLE; NO DELAY. An agent of the shipping company waited several days and when no one came to claim it, he sent the following postcard to Florida: "Please call for your Guavas. I think he is dead."

The akee is a dramatic-looking fruit with red skin, cream-colored flesh, and prominent external black pits.

ACORNS

> "That age of innocence, acorns and happiness."
> Writings of Moses

Acorns were representative of the idealized Golden Age when man could subsist on such simple fare as this fruit of the great oak tree. The Swiss cave dwellers of prehistory ate acorns; so did the ancient Greeks (the oak was often referred to as Jove's tree and acorns were called Jove's nuts). Because it takes twenty years before an oak produces, the nut often symbolized perfect achievement requiring long-term effort. An old proverb goes "Acorns were good until bread was found," and in time during the medieval period the nuts were relegated to the position of a famine food. The Pilgrims observed how the American Indians roasted acorns, ground them into meal, and mixed them with corn to make bread. Acorns from large white oaks were thoroughly washed, to eliminate any of the bitterness caused by the tannin in their skins, and then fried in oil. Sweet acorns were often eaten raw. Acorns are closely related to chestnuts and even today are roasted in the same manner in Spain, Portugal, and North Africa. In Turkey, a dish known as *racahout* is made from acorns that have been buried in the earth, then washed, ground, dried, and combined with sugar and spices. Elsewhere, they are rarely eaten. There are some poisonous species of acorns that can be harmful to cattle and pigs; the latter, like Piglet in *Winnie the Pooh,* love "haycorns."

The Celtic druids were said to have lived on acorns and have eaten them before prophesying. In fact, the name *druid* derived from the Greek word for tree (particularly oak), *drus;* a wood nymph was a *druas.* In Modern English, *acorn* developed through folk etymology from the combination of the Old English word for oak *āc,* and *corn,* literally meaning "fruit of the field" and closely related to the word *acre.*

ALMONDS

> "... And it came to pass, that on the morrow Moses went into the tabernacle of witness; and, behold, the rod of Aaron ... was budded, and brought forth buds, and bloomed blossoms, and yielded almonds."
> Numbers 17:8

Aaron's revered almond rod was said to have reached Rome, where it became the staff of the pope; and in the Tannhäuser legend, it became the symbol of forgiveness when it miraculously greened. Almonds have grown abundantly in the Holy Land since biblical times and were used as models for decorations on the candlesticks in the Temple, as well as on currency during the time of the Maccabees. In Tuscany, almond branches were used to determine the whereabouts of hidden treasure. Attis, one of the Phrygian gods, supposedly was conceived when his mother placed a ripe almond in her bosom.

In Israel the beautiful pink or white flowers of the almond tree appear even before the leaves do—as early as January. This phenomenon has made the almond tree symbolic; in Asian mythology it is often portrayed as the father of everything in nature; it can also signify a sudden or quick action, as recorded in the Bible. ". . . the word of the Lord came unto me, saying, Jeremiah, what seest thou? And I said, I see a rod of an almond tree. Then said the Lord unto me, Thou hast well seen: for I will hasten my word to perform it" (Jeremiah 1:11–12).

Almonds, probably the oldest and most widely grown and eaten of all nuts, grow in most temperate fruit-growing regions. They are native to North Africa and western Asia, but were cultivated throughout the Mediterranean region (the Romans called them "Greek nuts"). Almond trees grow from twenty to thirty feet high and are erect when young, spreading as they mature. The leaves of an almond tree are pale green with a gray tinge. The tree is a close relative of the peach and the nut is almost identical to the peach stone. The fruit of the almond tree resembles a wrinkled, flat dwarf green peach. In some European countries, green almonds are marketed. The fruit is not eaten but the immature soft pit is served with apéritifs or as dessert. Most often, though, the fruit is allowed to ripen, at which time the hull splits open and the mature almond is picked.

The unshelled almond resembles a peach pit.

Almonds fall into two general categories: sweet and bitter. Sweet almonds are those usually eaten alone or in combination with other foods. They are marketed with their orange-brown peel intact or blanched creamy white. The Jordan almond, one of the most prized sweet almonds, is not grown on the banks of the Jordan as one might expect, but in Spain. In fact, Spain and Italy are the major producers of almonds. In tropical regions such as Java and Malaysia, "true" almonds do not grow, but there are many nuts that closely resemble them and are given the same name. An edible oil, an extract for flavoring, and a high-protein flour (particularly good for people who must restrict starch in their diets) are made from sweet almonds. Bitter almonds are inedible, as are the peach kernels they resemble. They contain some of the same oil as the sweet variety but also contain a glucoside substance, which, when combined with water, yields prussic (hydrocyanic) acid and the essential oil of bitter almonds (benzaldehyde), both of which account for their extremely bitter taste. Prussic acid is poisonous but is removed by heat, and bitter almond extracts, like sweet ones, are used in baking and as costly bases in cosmetics (apricot- and peach-kernel oils are less expensive substitutes). Sometimes the oil is used as a mild laxative, and old herbals recommended it as a "purge" for children. Other therapeutic uses include those for insomnia and dysentery; as an antidote for the evil eye; a stimulant for lactation; a headache and hangover cure (five almonds eaten before drinking were supposed to prevent drunkenness). Pills containing extracts of liver, violet oil, and almonds were recommended for travelers in places where food and water were scarce in the sixteenth century. An eighteenth-century American Indian remedy for deafness was a concoction of red onions and oil of roasted almonds. The psychic Edgar Cayce suggested to some of his "patients" the eating of a few almonds each day to possibly inhibit a tendency to cancer; and in some southern states there is a belief that eating almonds can prevent and even cure cancer. (Apricot kernels, too, contain these cyanogenic glucosides.) Some people in the health field claim that, because the skin of almonds contains tannic acid, the nuts should only be eaten after blanching.

The English name *almond* comes from the French *amande,* which comes from the Latin *amygdala.* The Arabic prefix *al-* is seen in the Spanish *almendra.*

BEECHNUTS

The fruit of the beech tree is also known as beech mast and is eaten more by wildlife and domestic poultry than by humans. In times of famine, man has eaten beechnuts but, more often, it is the extracted oil that is used for illumination and cooking (because it keeps well). The nuts can be slightly poisonous, but the poison is water-soluble.

The words *beech* and *book* derive from the same Anglo-Saxon root, probably because the ancient Germans and Saxons wrote on thin slabs of beechwood. Beeches, in the same family as oaks and chestnuts, are valued for their wood and as shade trees. The American beech grows from seventy to one hundred feet high and the European often over one hundred feet high. The beech was also known as buck in England, and the county of Buckingham was so named for its thick beech forests.

BRAZIL NUTS

Brazil nuts do grow in Brazil but were not popular there until fairly recently. These deliciously rich large dessert nuts, a favorite of Europeans and Americans (who import more than half the entire South American production) usually come to the table either in their hard triangular shells or shelled and white, giving no indication of their fascinating trip from their beginnings as seeds of the Brazil nut tree. Actually, the Brazil nut that we eat is one of from twelve to twenty-five hard seeds, which fit so perfectly into a large (two- to four-pound) woody nut similar to a coconut that, once this is opened, it is impossible to put them back into the same space. This outer shell, called *ourico* in Portuguese, is used for cups and other utensils. Brazil nut trees grow up to 150 feet high, have no branches until at least 40 feet from the ground, and have trunks that may be six feet in diameter. The millions that grow in Brazil are not cultivated; instead they are propagated by hares who gather the nuts and bury some of them, a few of which take root. The tree blossoms like a hydrangea and takes about nine years to produce nuts. When the nuts ripen, human hands do not pick them. In a remarkable natural joint venture, strong winds snap the branches on which the nuts are attached and cause them to fall to the ground with such force that sometimes the nuts are burried by the impact and must be dug out of the ground. Naturally, anyone hanging around a Brazil nut tree at harvesting time is advised to wear strong protective headgear or suffer the serious consequences.

Brazil nuts are sometimes called para nuts, cream nuts, butternuts (an entirely different species), or monkey pots, the last-named because sometimes monkeys thrust their hands into the aperture at the end of the large shell (closed by a "plug" that can be cut or broken open) to try to retrieve the nuts. Brazil nut trees also grow in some other parts of South America, including Venezuela, but it is difficult to grow them elsewhere. Fat and oil are extracted from the nuts, which have the highest fat content (66 percent), and they are used in medicinal preparations, for lubrication, and for lighting. The husk of Brazil nuts is used for caulking ships.

Sapucaia or paradise nuts are related to Brazil nuts and have an almond-like taste.

BUTTERNUTS

Butternuts are also known as oilnuts and white walnuts and are similar to black walnuts. The tree differs from the black walnut in that it is usually smaller and has a lighter-colored bark, dark yellow, and wood with an attractive grain that is often used for furniture. The oval or elongated nuts, which are about two inches long, grow in clusters of from two to five. They have a hard husk, like other walnuts, and in the eastern United States in earlier times a brown dye was made from these husks for "butternut jeans." Butternuts are sweet, oily, and rich and are excellent in baked goods. Sometimes, while the shell is still soft and immature, they are boiled and pickled in vinegar.

Mentioned in *Potter's New Cyclopaedia of Medicinal Herbs and Preparations,* butternut bark was used by the American Indians as a remedy for worms and as an emetic and cathartic. The inner bark was boiled until thick and made into pills or syrup.

CASHEWS

Cashews grow in a most distinctive manner. A single kidney-shaped, olive-colored nut hangs beneath a bright-orange pear-shaped fruit three times its size. The fruit is called a cashew apple and is usually not eaten. It contains a substance that is reported to stain forever anything it touches and whose only use is in a fermented state as a liquor called *kajú*. One does not lightly go about picking cashews. The tree is a relative of poison ivy and sumac, and the nut shell has an oily substance so toxic and irritating that it must be burned off before one can touch it. Even the burning must be done with care; if the smoke comes into contact with any of the mucous membranes of the eyes, nose, or throat, it can cause extreme pain and even ulcerations. This lethal oil does serve a function, though: it is extracted and used for waterproofing, as an insecticide, and for other industrial purposes. The picking job is still not complete even after the burning off. The nuts must be boiled or roasted and a second shell removed before they can be eaten. Machinery has not been perfected for this arduous procedure and it must be done by hand. It is no wonder that cashews are so costly!

Cashew nuts grow on medium-sized trees native to the Central and South American tropics. The tree was one of the first fruit trees brought from America and planted in Asia. It thrives in drier areas than most nut trees and is related to the mango and pistachio. Cashew production is greatest in India, Mozambique, and Tanzania, and the United States imports the most nuts from these regions. Although there are "raw" (boiled) cashews, most of the marketed nuts are roasted. They are very rich: about 45 percent fat and 20 percent protein. The wood of the tree is used in the manufacture of shipping crates, boats, charcoal, and gum arabic. The Portuguese explorers are credited with bringing the tree to India in the sixteenth century and the name derives from the Portuguese *acajú*. In India the tree is known as *caju*.

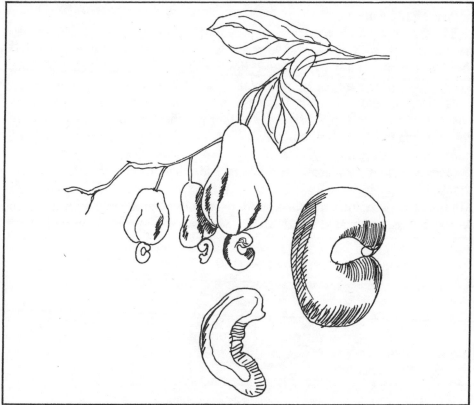

Cashews are appendages of an inedible fruit.

CHESTNUTS

> Under the spreading chestnut tree
> The village smithy stands.
>
> Henry Wadsworth Longfellow

Once the American chestnut grew abundantly from Maine to Michigan and from the Appalachian Mountains to Alabama. But most of the trees were destroyed by a fungus blight at the beginning of this century. Today many of the chestnut trees growing in the eastern part of the United States are hybrids of American, Japanese, and Chinese species, crossed to effect a stronger, disease-resistant strain; European or Spanish chestnuts grow on the Pacific coast. It is the latter type that produces the large chestnuts most commonly eaten raw, roasted, or boiled. Many of these chestnuts are imported into the United States from southern Europe and North Africa; the largest are known by the French as *marrons*. Chinese chestnuts, native to China and Korea, are sweet, and the Chinese sometimes make a kind of pasta from them. Sometimes chestnuts are fed to livestock, and chestnut-fed pork is highly valued.

The Greek Xenophon's army supposedly lived on chestnuts during its retreat from Asia Minor in 401–399 B.C. Chestnuts have a sacred connotation: they are eaten in Tuscany on Saint Simon's Day and on the Feast of Saint Martin, when they are distributed to the poor. Symbolically, chestnuts mean haughtiness or success to the Japanese, but to dream of them may portend business difficulties. In Christian symbolism, they denote chastity. More earthbound, the French idiom *travailler pour des marrons* means "to work hard for a small gain."

Chestnut trees can grow to be very old. One in Sicily grew for two thousand years until it was destroyed by the eruption of the volcano on Mount Etna. The trees are large, seventy to ninety feet high, with a crown sometimes as much as a hundred feet across. Also on Etna, five of them

Most chestnuts are eaten roasted or boiled as a vegetable.

standing together grew into one tree, with a trunk that measured seventy feet thick. Chestnuts are characterized by male catkins, which are golden in early summer, turning to light green in the fall. The female flowers are reddish. At fruition, two or three nuts are contained in the spiny burs that develop.

Chestnuts have been used medicinally to ward off rheumatism and to treat backache. Supposedly they were good for the blood but could make it too thick, which might in turn cause headaches if one overindulged. An old cure for asthma was an infusion of boiled chestnut leaves mixed with honey and glycerin.

Other families of chestnuts include the American chinquapin, whose burs each hold a single nut, and the horse chestnut. Horse chestnuts are often bitter, containing much tannin, but after soaking can be ground and mixed with other flours for baking. The California horse chestnut, sometimes called the buckeye because the brown nut has white eyes, is sweeter. The Moreton Bay chestnut grows in Australia. Its nuts grow in pods like peas and cannot be eaten fresh, but are dried and roasted.

HAZELNUTS (FILBERTS)

Hazelnuts and filberts are so closely related that they can be considered interchangeable. The name *filbert* is a corruption of the name of a Norman saint, Philibert, whose saint's day coincides with the time that the nuts ripen. Sometimes the nut is called full-beard, a reference to its hairy husk. The ancient name was *nux Phyllides,* the "nut of the nymph Phyllis," whom legend says was turned into a nut tree. The hazel participates in many old tales and superstitions; in Sweden, carrying a hazelnut in one's pocket could render one invisible; Circe used a hazel rod to turn her lovers into swine; and the first Christian church in Glastonbury, England, might have been built as a wattled house of hazels. In literature, Shakespeare likened a

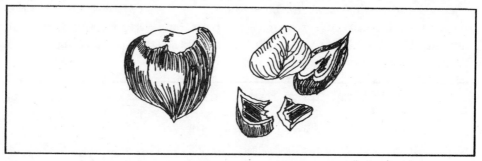

There are slight technical differences, but hazelnuts and filberts are usually treated the same.

few of his heroines to the hazel: "Sweetest nut has sourest rind, / Such a nut is Rosalind" (*As You Like It*); "Kate, like the hazel twig / is straight and slender, and as brown in hue / as hazelnuts and sweeter than the kernels" (*The Taming of the Shrew*). Forked branches from the hazel tree were long used as divining rods: the dowsers believed that the branch would twist if water, minerals, or treasure lay beneath the surface. The quality of the rod was tested by holding it in water; if it made a sound like the squeal of a pig, it was accurate. It is told that when Adam was expelled from paradise, God pitied him and allowed him to create new animals by striking water with a hazel rod. Hundreds of years ago, it was a custom in certain parts of France to take out all the kitchen utensils on Midsummer Eve and make an enormous racket by knocking them together. This noise was supposed to prevent rain from falling and damaging the hazelnuts.

Hazels were often symbols of love, and to woo someone under the tree was to ensure reciprocation. In France and Germany, certain expressions meaning "to make love" include the words for hazelnuts; young girls would dance under the tree to attract suitors; and the nuts were given to the bridal pair. There also is an association with the attribute of justice. Before the seventeenth century, hazel branches were employed to find thieves and murderers. In Prussia, the clothing of a suspected thief was beaten with hazel branches, and if he became sick as a result, this would prove his guilt. The hazel tree is prominent in Celtic legend, particularly Irish, where it is characterized as the tree of knowledge and linked with the magic number nine; the nine hazels of wisdom signify all knowledge of the arts and sciences; it is the ninth tree in the Old Irish tree alphabet, and a symbol of the ninth month (August 6 to September 2). So honored was it that anyone caught cutting down a hazel tree could be put to death. The Norse chose it to represent the evil aspect of knowledge.

The hazel is a small tree, reaching a maximum height of twenty feet. Catkins form, and then the nuts (which, like acorns, grow in little cups). These are a rich medium-brown color, which came to be known as hazel. When the nuts are fully ripe, the husks turn brown and the nuts fall down and will rattle when shaken. They should be gathered only at this time. American hazels mainly grow in Washington and Oregon and bear small nuts of little commercial value. The European hazel is more common. England, Turkey, and Spain are the leading producers. Wild hazelnuts have been eaten since earliest times and were recorded by Pliny. The hazel was the most common nut tree in Britain and formerly was simply known as the nut. In addition to being a nutritious eating and baking nut (sometimes also ground and cooked in meat dishes), hazels were mixed with mead, honey, and water as a remedy for persistent coughs or mixed with pepper to clear the head. Superstition made certain Britons carry a double nut in their pockets to prevent toothache. Hazelnut oil is used as a base for perfume, and fine drawing charcoal is made from the wood. Hazel trees start nutting at the age of six, reach the height of production at fifteen, and can continue producing until fifty.

HICKORY NUTS

Hickory trees are among the most important timber trees in the eastern region of North America. All hickories bear nuts. Most are edible, and taste is the guide. Of the many species of hickory, the pecan is the best-known nut-producer (see *Pecans*). Second in importance is the shagbark hickory, so named for its loose, shaggy bark hanging in long strips. Though fewer shagbarks are cultivated, the nut, with its thin, flattened, light-tan shell, is the northern equivalent of the pecan. Shagbarks grow very old; some of them have been found to be more than 350 years old and measure as much as four feet across. Another closely related hickory is the large (seventy- to ninety-feet) shellbark, which produces similar nuts. Hickory nuts can be eaten raw but are especially good when baked in cookies. The hard wood of the hickory is valued in the manufacture of tool handles, lawn furniture, and skis, and for smoking meats and as fuel. The name *hickory* has its origin in an American Indian nut milk called *powhicora,* which was made from pounding the nuts and shells together and then boiling them in water.

MACADAMIA NUTS

If you fly back and forth to Hawaii on a regular basis, you can support the macadamia nut habit free (samples are given to passengers). Otherwise, it will take a lot of money. Macadamia nuts are those rich dessert and cocktail nuts usually sold in jars, fully processed (shelled, graded, roasted, and salted). Although native to Australia, where they are known as Queensland nuts, they are imported here almost exclusively from Hawaii. Cultivation had not been possible on the American continent. Macadamias grow on a large ornamental tree, twenty-five to thirty feet high and fifteen to twenty feet wide, whose long, rectangular leaves stand out almost perpendicularly from the branches. The nuts are spherical, are encased in a hard brown shell, grow in clusters, and are allowed to fall instead of being picked. It is doubtful that many of us will see a macadamia nut in its virgin state, but to the aborigines of Australia it is an important dietary staple, and no doubt its high fat content sustains them well.

PEANUTS

Peanuts are really legumes, like peas, but are more often eaten as nuts than as vegetables. Peanut seeds have been found in ancient Peruvian tombs, and peanuts are believed to have originated in South America, although there are some who believe Africa to be their original habitat. In any case, the nuts, also called groundnuts, earthnuts, and goobers, were cultivated by the Aztecs and Mayas, and the Spanish explorers brought them to Spain and Africa. In the African slave ships peanuts were a major source of sustenance, but, once in America, they fell into disrepute with slaves and masters alike and were cultivated in the South only as a floral curiosity and for hogs (peanut vines are as good as clover hay).

"I could either sit down and cry over it or I could improve it." This is how George Washington Carver summed up the soil he saw at Tuskegee, Alabama, when he arrived there in 1896. The South had suffered seriously from being a one-crop region, and cotton had depleted the land of those nutrients required for healthy plant growth (and for well-fed people). A former slave, Carver was a genius who embodied science and spirituality, exemplifying that the two supposed opposites could work harmoniously together. One of his many accomplishments was the development of a new science, the study of soil constituents in order to increase and improve production, and the land his people were struggling with presented him with a singular challenge. The soil had to be enriched, particularly with nitrogen, and legumes were excellent for this purpose. Why not try peanuts, which were an inexpensive and accessible legume? Carver chose a dramatic way to prove his theory: he planted peanuts on twenty acres of the worst Alabama soil, land that had yielded a loss of sixteen dollars an acre. After the first year, there was a profit of four dollars an acre, and after seven years of rotation with cotton, a profit of seventy-five dollars! But the result was far greater than economic. Peanuts are extremely nutritious; high in protein, containing much iron and vitamins. A pound of peanuts has the same or more body-building nutrients than a pound of sirloin, and twice as many calories. It could be said that Carver not only helped to create a new agricultural industry (peanuts are one of the leading crops of

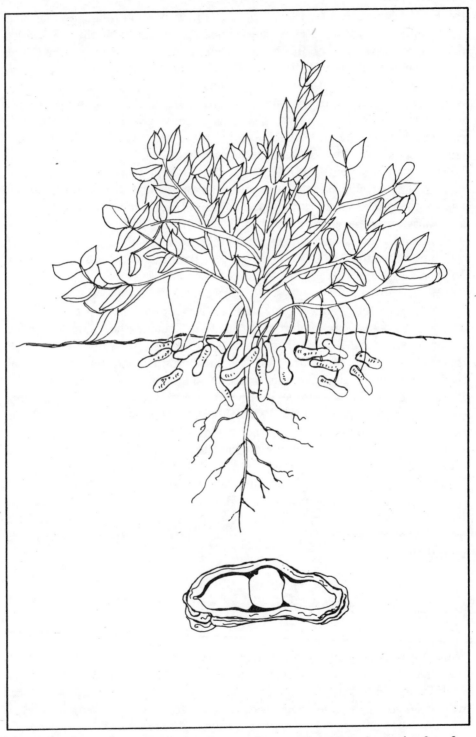

Bunches of 25 to 50 peanut pods mature underground; each pod contains 2 to 3 (Virginia) or 3 to 4 (Valencia) nuts.

the United States, representing a production of over 1 billion pounds and $200 million a year) but also transformed the eating habits of the South's poor. This "inventor" of the peanut then found over three hundred uses for it—peanut oil, peanut brittle, and peanut butter are the most common, but some other items derived from peanuts are: flour, shaving lotion, soap, dyes, a coffee substitute, imitation cheese, material for insulating boards, and even a dandruff treatment. The higher-quality grades of peanut oil are excellent for cooking and salads, and the lesser grades for packing sardines and other foods. The oil has been used medicinally for a long time. Because of its adhesive ability, it has been added to liniments for polio victims. The psychic Edgar Cayce often recommended it be rubbed on the base of the spine for arthritic pain. The peanut has a function in all stages: after harvesting, the plant is used as livestock feed, and peanut shells have such diverse uses as fuel, polish for metal, and linoleum manufacturing. If you buy peanuts in the shell, after you eat the nuts feed the shells to your house plants. They're a great soil enricher.

Peanuts grow in two ways: on bushes (up to two feet) with short branches or on long runners lying close to the ground. The first type, known as "bunch" plants, are preferred by growers because they can be seeded closer together and harvested more easily. In both forms, after pollination the stalk with the immature pod becomes rootlike and burrows several inches into the ground. For this reason a loose, nonacid soil is necessary. Peanuts grow best where the weather is warm for at least five months of the year. A light, sandy soil, which produces lighter-colored shells, is desirable for peanuts sold in the shell (which can last as long as two years if kept covered and cool), although shelled peanuts are more popular and are available roasted and salted or raw. The best-known varieties in the United States are the large Virginia (used mostly for roasting) and the smaller Spanish and Valencias (used almost exclusively for peanut butter). Harvesting is difficult and usually done by machine, and the nuts, still attached to the vine, are allowed to dry in a warm, airy place before being picked and sorted.

PECANS

Some North American Indians believed the pecan tree to be the manifestation of the Great Spirit and valued it so much that the early Spaniards in Florida were able to trade nuts for hides and mats. One tribe, the Mariames, who lived in Texas, ate them as their only food for two months of the year. The Indians made a meal from the ground dried nuts and sometimes would chew the nuts to a mush to force-feed starving infants. In fact, the name *pecan* is from Indian sources—Cree *pakan,* Algonquian *paccan;* all nuts with hard shells, such as hickory, were called *pâcan* by the Indians in the Louisiana region, and the French who settled near the Mississippi basin used the name for a particular species found growing wild and abundant in Louisiana. The pecan was appreciated not only for its nuts; the tree, sometimes growing to a height of 180 feet and a diameter of eight feet, shaded with unusual grace; and its deeply furrowed grayish to reddish bark sheathed an excellent wood used for heavy furniture and flooring. Pecans grow best in river bottoms and lowlands and are sensitive to frost. They grow wild throughout the Mississippi Valley region and the river valleys of Texas. They were first cultivated in El Paso at the end of the seventeenth century, and the largest-producing cultivated areas are in the South. (After the Civil War, Union soldiers brought pecans home in their pockets and attempted to plant them farther north.) Cultivation has been extended in recent years eastward and westward. Commercial pecan production has grown since the beginning of the twentieth century and now pecans are the first in economic importance of the native North American nut trees and the fifth leading tree nuts of the world.

Traditionally, pecan trees are hit with bamboo canes to bring the nuts down. The best quality are thin-shelled and easy to crack. Most commercial varieties are olive- or oblong-shaped and have polished shells; some are dyed red (the natural color is brown). Pecans are eaten raw, ground into a digestible meal, and baked in pies, and they are the main ingredient, with maple syrup, of pralines, a favorite candy of the South. Possibly because of some indication of a concentration of B vitamins, pecans are sometimes recommended in the treatment of arthritis.

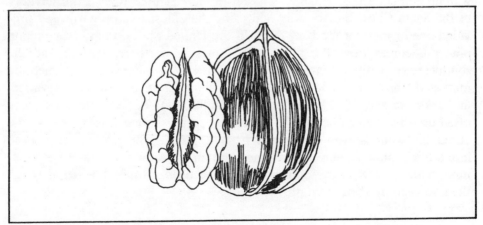

The smooth brown shell of the pecan is often polished.

PIGNOLI (PINE NUTS)

Pignoli (singular: pignolia) nuts have been harvested from pine trees in dry southern European countries since prehistoric times. The most popular edible seeds of pinecones are those from the Mediterranean stone pine, which have been relished since the time of ancient Rome. The stone pine trees rises to a height of eighty feet, is umbrella-shaped, and bears glossy brown round cones. These cones must be exposed to the sun in order for the scales to open and the seeds to be extracted. Pignoli are luxury items, mostly because of the difficulty in extracting the very small (half-inch) nuts; the pinecone is filled with a sticky resinous substance and each of the many hard-covered nuts must be shelled by hand. Unshelled pignoli are rarely sold. Most often, they are found already shelled, with a slight pine taste, and very rich both in texture and price. The demand far exceeds the supply, as the crops are usually uncertain and therefore pignoli are generally used sparingly for cooking rather than for nibbling. The oil from pignoli nuts is sometimes used for cooking and illumination.

In China and Japan the pine is symbolic of longevity and the Chinese god of immortality is often pictured sitting beneath the tree. The cone is also a Semitic symbol of life.

Nut-producing pines grow in other regions, such as Chile, where one, known as the monkey puzzle, bears cones the size of a man's head, containing from 100 to 200 *piñones* ("nuts"), which are roasted and eaten as food. The aborigines of Queensland, New Zealand, feast on the large nuts of the bunya bunya pine and reportedly grow fat on them because of their high starch content. In North America there are about thirty species of nut-bearing pines growing from Quebec to Florida; one has nuts the size of walnuts, others are as small as a grain of rice. A pine tree in the territory of the Santa Clara Indians was said to be the oldest tree on earth and provided one of the first foods of man. In the United States, the most common pine nut comes from the piñon and grows in northern Mexico and the southwestern United States. Its nuts, known as piñons, are sometimes marketed simply as "pine nuts." They are similar to the European variety and easier to extract (like the so-called India nut). Boiled pine cones were often used by the old herbalists (a fourth-century botanist noted a fox who cured his wounds with droppings from a pine). The nuts were pressed into a thick, high-protein milk. Pignoli contain more protein than most other nuts; a fact the Roman soldiers must have been aware of when they ate them to keep up their strength.

PISTACHIOS

"... take the best fruits in the land in your vessels, and carry down the man a present, a little balm, and a little honey, spices, and myrrh, nuts, and almonds."

Genesis 43:11

The nuts were pistachios that Jacob's sons carried to Egypt. The pistachio originated in Syria and has grown in the Mediterranean region for at least five thousand years; the village of Betonim mentioned in the Book of Joshua received its name from the abundance of pistachio trees growing there. Pistachios were a favorite of the Roman artistocracy, who carved toothpicks from the thick, leathery leaves of the tree. One gluttonous emperor of Rome, Vitellius, would finish off his meal by stuffing his mouth full of pistachios.

The pistachio tree once grew wild in parts of Russia, Turkestan, and Afghanistan and is now cultivated in areas of the Middle East and the Mediterranean region; the leading producers of the nuts are Iran, Turkey, and India. Although there is some cultivation in California and Oregon, most pistachios are imported and consequently high-priced. The pistachio is a broad, bushy tree that usually grows no more than thirty feet high and sometimes has more than one trunk. Its bark is russet; its wing-shaped leaves are gray-green and, when bruised, give off the same aroma as the fruit, which grows in thick clusters. Inside the reddish, wrinkled fruit is the hard, double-shelled, oval nut. Pistachios are naturally white, a fact most Americans may be unaware of since invariably they are sold covered with a red dye that "bleeds" on everything when wet (they can also be dyed green or blue!). But natural, undyed pistachio nuts, plain or salted, can be obtained at some natural-food or Middle Eastern specialty stores and are well worth seeking out. Pistachio shells are usually split at one end, which may tempt the eater to open them with his teeth or nails. Although this is usually effective, it is risky and may lead to more broken teeth and nails than shells. A more prudent way is to use half a shell from an already opened nut to pry open the closed one. Once opened, the nut will be seen to have a dry, thin, reddish covering over a pale-green, rich, and delicious kernel. Pistachios are sold roasted and salted, sometimes shelled. They star in such confections as ice cream, Indian halvah, and Turkish delight. For a more unusual taste, try adding them when cooking dishes like poultry stuffings or when blending milk drinks.

WALNUTS

"I went down into the garden of nuts to see the fruits of the valley. . . ."

The Song of Solomon 6:11

The walnut was among Solomon's most prized trees. In Christ's time it grew along the shores of the Galilee and the nuts produced the brown dye for his coat. The Greeks called it the Persian tree (it is native to Iran) and believed that, while men lived on acorns, the gods feasted on walnuts. Hence it was the royal nut and dedicated to either Diana or Jove. In Rome, walnuts were thrown (hopefully, unshelled!) at the bridal couple as rice is today; they were served, stewed, at the wedding feast; and it was customary for the groom to strew the bridal chamber with walnuts as an offering to Jupiter to grant him patronage and bestow the virtues of Juno upon his bride. The next day, children were invited in to pick up the nuts. Obviously a symbol of fertility, the walnut in this context was also interpreted as the assumption of the responsibilities of adults by the married couple who were putting away the frivolities of youth (walnuts were children's playthings).

"A dog, a wife, and a walnut tree: the more you beat them, the better they be."

Old saying

In time, walnut trees took on another significance; to dream of one augured danger, misfortune, or unfaithfulness. European folklore is replete with beliefs that witches and other vile spirits hide in the branches or gather under the tree (but dropping a nut under a witch's chair will permanently root her to the spot). In some countries it was customary to beat the trees, supposedly to make them yield better but more possibly an offshoot of the superstitions. In Bavaria a half-burned branch was used to ward off lightning (Jove's thunderbolts). As previously noted, the walnut husk was used as a dye, not only for fabrics but also for the skin to create a disguise (the freshly hulled black walnut, which grows throughout the United States, can leave a brown stain virtually impossible to remove) and for the hair (a poultice of damp hulls left on the head overnight could remove all traces of gray). In accord with the herbalists' doctrine of signatures (which purported that the appearance of a growing object was a clue to the part of the anatomy that it could treat), the walnut, because it resembled the head, was used for ailments located there. The walnut's two-lobed kernel looks like the convolutions of the brain, a point noted by the Greeks who called it *kara* ("head"). The walnut was, therefore, used to treat all kinds of mental illness and head injuries and deformities. Nicholas Culpeper referred to it as the tree of the sun and recommended using it while still green (before the shell developed) as medicine for toothache and deafness; the oil for colic; the mature nut meat for carbuncles and (in red wine) for falling hair; the leaves distilled in water for running sores because they were a powerful astringent. This quality of astringency became the bane of farmers, who claimed the leaves so astringent as to be detrimental and almost poisonous to any vegetation on which they might fall.

Walnut oil, which is edible, has a special function in painting; it was used extensively by such master painters as Van Eyck and Correggio because it dried faster than any of the other media available at that time and facilitated the application of different layers of paint.

There are several species of walnut trees; the two most important are the English and the Eastern black. English walnuts are the aforementioned Persians, which grow in many European countries (abundantly in France) and on the West Coast of the United States. The tree grows very tall (up to a hundred feet) and requires high humidity and little temperature change to thrive. Both the English and the black walnut are prized for their wood and used in the manufacture of fine furniture. The beautifully grained Circassian walnut used for veneers comes from English walnut trees growing in France and parts of southern Europe. Most commercial production of walnuts is from the English species. The hard-shelled oval or round nuts develop from catkins in clusters of three or four. Husks open in the fall and nuts drop to the ground (a process that may be hastened by knocking the tree). The nuts should be picked immediately and dried in a single layer in a shaded, airy place. Commercial production is mechanized and walnuts are sold shelled or unshelled; unshelled, they have a covering thin enough to break with an ordinary nutcracker. (I have seen someone throw walnuts against a window and crack them without so much as a splinter in the glass, but the reason is a mystery to me and I strongly advise against it unless you are thoroughly knowledgeable.) "Green" (immature) walnuts are often pickled in vinegar, preserved in syrup, or made into a liqueur (*brou de noix*); in the Middle East, walnuts are sometimes sold husked and soaked in cold water so that the skin can be peeled off easily, and in Russia, a delicacy is candied walnuts on a stick.

The black walnut is a native of North America and grows best where the soil is moist and rich, such as New England, Florida, Nebraska, and eastern Texas. It also is a tall tree, with a dark-brown, deeply grooved bark. The heartwood of the black walnut is considered among the finest for furniture and interior finishes; it has a hard, straight grain that does not warp or split, is easy to work on, and will take a high polish easily. Black walnut wood was early exported to Europe by the American colonists. The nuts, rich and oily, are favored for use in candies, ice cream, and baking goods. Wild black walnut trees produce nuts which can be used for the same purposes but which have such hard shells that a hammer is more in order than a nutcracker.

MISCELLANEOUS NUTS

Betel Nuts, also known as areca from the palm tree on which they grow, are chewed, together with their shells, like tobacco in their native Eastern countries and in the Philippines. The tough nut is an astringent that cleanses the teeth (the teeth become black in the process and will remain so with too much betel-nut chewing), deodorizes the breath, and helps digestion. Chewing betel nuts causes one to spit a red fluid and get a bit "high" if not used to the habit. Often flavored with lime, betel nuts are eaten at puberty, marriage, and death ceremonies, and there is much elaborate equipment (like Western smokers') to accompany the chewing. Leaves can also be chewed.

Breadnuts are related to the breadfruit. The nut is a single seed, which is boiled or roasted. It is nutritious and tastes a bit like a hazelnut. Breadnuts grow in Jamaica and Mexico.

Candlenuts are so called because they are so rich in oil that, when strung together, they can burn like candles. The stone of a large fruit grown in tropical Asia and the Pacific, the nut is also called *bancoul,* is very nutritious, and tastes like a walnut.

Clearing Nuts are not edible but are named for their valuable function of purifying water. The cut seed is rubbed on the inside of the water container and impurities sink to the bottom.

Fox Nuts have been cultivated in China for more than three thousand years. A fox nut is the ornamental black seed of a kind of water lily that is roasted by placing it in hot sand. Fox nuts are popular in India and used like arrowroot there.

Ginkgo (Ginko) Nuts are those oval, mealy, yellow items that you think are beans when you find them in a Japanese or Chinese dish. Roasted ginkgo nuts are often eaten at oriental weddings and other feasts as a dessert nut to aid digestion and counteract drunkenness. The nut grows on the maidenhair tree in China and Japan and is the stone of a small, inedible, foul-smelling yellow fruit.

Kola Nuts are cultivated in the American tropics and in Africa, where they are chewed as a stimulant. They are used medicinally as a diuretic, for diarrhea, and as a heart tonic. They contain caffeine and are the base for many cola drinks.

Other nuts known for uses aside from eating are cohune, babassu, dika, and grugru for oil and soap-making; tonka for perfume and synthetic vanilla; tung for paints and varnishes; and tagua or coroza for buttons.

SEEDS

In the seed of all plants is a natural store of the nutrients and enzymes necessary for healthy growth. Seeds are valuable sources of vitamins, minerals (especially magnesium and the rarely found trace minerals), amino acids, and lecithin. Most seeds are a source of complete protein and generally are more digestible and less acid than animal protein. Although many Westerners do not eat plain seeds as part of their daily diet, vegetarians in parts of Asia, including the Middle East, have been using them for centuries. When one considers the benefits, seeds are hardly just "for the birds." Some seeds can be eaten plain or roasted like nuts, some are more important for the nutritious, unsaturated oil they yield, and others are excellent for sprouting. The seeds of vegetables and fruit, such as squash, pumpkin, and watermelon, have been discussed under those headings. What follows is a brief description of some other edible seeds and ways of using them.

Alfalfa is a legume like beans and peas, but the leaves and stems can be eaten in addition to the seeds. Alfalfa has been around for thousands of years, but for most of that time the cattle were its beneficiaries. Alfalfa plants burrow deep into the ground (ten to twenty feet) and thereby develop high proportions of calcium, iron, protein, and vitamins, especially A, E, and K. Alfalfa seeds are easy to sprout and provide a fresh salad green year round, but most of the nutrition is still maintained even after drying the leaves and stems for infusion into tea.

Anise seeds were eaten by the Romans to aid indigestion and even today are an excellent natural cure for stomachaches and dysentery. Ground into powder, they can be added to baked goods or yogurt, and a liqueur called anisette is derived from them.

Carob is colloquially known as Saint John's bread, in the belief that it was the "locust" recorded in the Bible as being eaten by John the Baptist in the wilderness. Actually, the pod and not the seed of the carob tree is eaten and ground into a powder that is often sold as a more wholesome substitute for chocolate.

Cumin seeds are generally used as a spice in Indian curry and Middle Eastern dishes, but are also medicinal as a stimulant and carminative against colic and flatulence. They are very hot and are not to be eaten straight.

Dill seeds are an important ingredient in pickling (viz., dill pickles) and seasoning vegetables and salads. In the nineteenth century they were infused into a tea to cure obesity.

Flax seed is rarely eaten plain. The plant grew wild in the United States and the seed produced an oil used in paints and varnishes. Flax seeds are laxative and can be infused into tea for this purpose. In ancient Rome they were a great delicacy and were also used to treat respiratory ailments.

Lotus seeds are not available here; they are the seeds of the flower. Because of their high content of calcium, phosphorus, and iron it is logical that they would give energy to the fabled lotus-eaters.

Poppy seeds are so soporific (although opium does not come from the seed of the poppy) that the name actually derives from the custom of mixing them with children's pap to put them to sleep. They can be chewed or infused as a natural sleeping "pill," and in small quantities season baked goods and noodles.

Safflower seeds yield one of the most unsaturated, lecithin-rich oils available. The seeds are not eaten plain, but were once used to dye articles red in their places of origin, India and China. Safflower oil is preferred as one of the fastest-drying media for oils by painters.

Sesame Seeds The expression "Open Sesame" in oriental folklore denoted an entry to splendor and richness, and the knowledge the ancients had of the value of sesame seeds probably underlay it. The seed was used by them and became known in Egypt and the Near East about 1300 B.C. Sesame is perhaps the most widely used seed and has been eaten in a variety of ways for

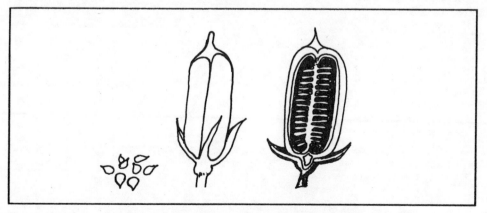

Sesame seeds are used in making some candies, notably halvah.

hundreds of years, primarily in Turkey and the Middle East. It is extremely versatile; one can eat the tiny sesame seeds plain and raw (the unhulled variety that are white or tan are more nutritious and preferable to the hulled seeds), but they taste better lightly toasted in a pan and mixed with vegetables or ground with sea salt into a tasty seasoning called gomasia, which can be made at home or purchased in natural-food stores. Sesame seeds are deliciously present in halvah and other similar confections, tahini (sesame paste), and sesame oil, which may be used for dressings, for frying, and as a base for shampoos, soaps, and cosmetics. Sesame oil is also used for a luxurious body massage.

Sunflower seeds are chewed extensively by the Russians (at least, that's what their novelists would have us believe) and those who chew them are supposed to have excellent teeth, because the seeds are exceptionally high in vitamin D, as well as the B complex and E vitamins. This makes them a good antihemorrhagic, particularly for pyorrhea. For a long time, sunflower seeds were called polly seeds and relegated to feeding parrots. Sunflowers grow in temperate regions and some can reach such heights that it is claimed that a man on horse can stay hidden in a field of them. This flower, which Van Gogh immortalized, is attractive to bees as well and becomes a source of nutritious honey. The seeds are really dry fruits and are white, brown, black, or black-and-white-striped. The seeds can be eaten raw or roasted, shelled or unshelled, and can become habit-forming (nutritious, luckily). Meal is ground and oil also extracted from the seeds. Supposedly sunflower seeds were introduced into Europe from the area of New Mexico by the Spanish in the beginning of the sixteenth century, and Peter the Great is credited with bringing them to Russia.

Preparation of Vegetables and Fruits

Losses of carbohydrates and vitamins (particularly vitamin C, most of which is lost in green leaves within two or three days) begin as soon as a plant is picked and increase during transportation and storage. Buy only fresh produce and, if refrigeration is necessary, don't wash it beforehand; wrap it well and use it as soon as possible. "Winter" fruits, like apples and pears, and certain root vegetables may be stored for longer periods where there are correct conditions of temperature and humidity.

Regardless of how they are to be eaten, all fruits and vegetables should be carefully washed before using. Those grown in compost without chemical or animal manure will not usually have worms, but even organically grown produce may have residues of insecticides or harmful chemical sprays.

A stiff natural-bristle brush will remove most surface dirt; washing should be done in a solution of water and salt or vinegar, and the most exacting method is to allow produce to soak for a period of time (although this does contribute to vitamin loss). Particular attention should be paid to cleaning around stems and blossom ends, because these seem to be the areas of greatest accumulation of sprays. Fresh juices can be cleaned of bacteria by adding lemon juice (1 part lemon juice to 4 parts other fruit juice). Of course, skins can be peeled* but, whenever possible, try to avoid this—a lot of nutrients are contained in the skins of vegetables and fruit. In fact, try to use as much of the plant as possible: leaves, stems, seeds, etc.

The best and, for me, tastiest way to cook vegetables and still conserve their valuable nutrients is to use as little water as possible. Two methods are recommended:

(1) Steaming, in either a special steamer, a pressure cooker, or a small stainless-steel basket set in a heavy pan with about a half inch of water. The pan should be almost covered, with just a little "breathing" space, and no salt used while steaming. The water should not be discarded, since it contains valuable vitamins, but should be used in sauces, soups, etc. (2)

---

* Skins should be peeled when they have been waxed (an unfortunate modern practice). No amount of scrubbing or strong solutions will remove wax.

Sautéing quickly, in a heavy pan with a small amount of oil over high heat for about five minutes; then covering tightly and simmering until tender. This method is ideal because the vegetable cooks in its own juice and nothing is lost. If necessary, a little boiling water or consommé can be added while simmering.

When the terms *cook* or *sauté* are mentioned in the glossary, they apply to the methods described above. Pots and utensils used for vegetables should be made of stainless steel, heavy iron (plain or glazed), or glass. Aluminum should be avoided, especially for foods containing acid.

Vegetable Glossary

ARTICHOKES

Consumer information Available fresh from March to June. Available all year canned, bottled, or frozen. Look for compact, heavy, plump globes that have large, tightly clinging, fleshy leaf scales. Artichokes should have a good green color.

Storing Keep in refrigerator and use within a few days.

Uses Whole artichokes may be steamed or baked (plain or stuffed), served hot or cold. Artichoke hearts may be deep-fried in batter or marinated for salads. When using whole, snip off points of leaves with kitchen scissors or sharp knife. Trim artichokes at base (stem end) and rub with lemon to prevent darkening before cooking. Force open center leaves from the top and cut out choke (if desired). An artichoke is cooked if leaves come off easily at a slight pull. Remove from liquid and drain by standing stem side up.

ASPARAGUS

Consumer information Available fresh from March through June. Available all year canned or frozen. Look for stalks with the largest amount of green, because only the green portion of fresh asparagus is tender. Stalks should be firm and fresh and have compact closed tips. Open tops are a sign of overripeness. Avoid angular or flat stalks, which may be woody. Asparagus should be kept cold in the store; if kept at room temperature, it tends to turn fibrous. Avoid asparagus with stems soaking in water or which appears to have been soaked.

Storing   Keep in refrigerator and use as soon as possible.

Uses   Cut off about 1 inch from bottom of stems. Asparagus may be steamed, baked (plain, with sauce, or au gratin), or marinated. Serve hot or cold, the latter in salads and sandwiches. Raw asparagus, cut in small pieces, may be used in salads.

Hint   Steam asparagus, tied together in a bundle of 8 or 10, standing upright in about 1 inch of water in a pan with a tight cover. Before serving, drain and roll carefully in a dish towel to remove all excess water.

BEANS, DRIED

Consumer information   Available all year.

Storing   Dried beans will last indefinitely. Store in a tightly closed container.

Uses   Most dried beans should be washed well and can be baked without presoaking with the addition of a bit of oil and boiling water. If presoaking is desired, boil for a few minutes beforehand to hasten the process. Do not use baking soda! Beans can be boiled or baked. After cooking, they can be fried, mashed, etc. Serve them hot or cold in main courses, soups, casseroles, pies, and salads.

BEANS, GREEN

Consumer information   Available fresh mostly from May to August. Available all year canned or frozen. Green beans should be free from scars and discolorations and be fresh in appearance. They should break with a snap and have no strings.

Storing   Keep in refrigerator and use as soon as possible.

Uses   Green beans may be steamed, sautéed (particularly good this way with almonds), or marinated. Use in salads, raw or cooked.

BEETS

Consumer information   Available fresh all year, with peak June through October. Available all year canned, in jars (plain or pickled), or frozen. Early beets are often sold in bunches with their tops intact. These greens are usable if fresh and unblemished (see *Spinach*). Select small or medium-sized roots that are firm and have a good deep red color. Shriveled beets are undesirable.

Storing   Keep in refrigerator and use as soon as possible.

Uses   Beet greens may be used as a vegetable, in the same manner as spinach, but should be crisp and fresh. Beet roots may be steamed, baked, fried, or pickled. Serve hot or cold; the latter in salads and borscht. Slice raw beet roots thin or grate for use in salads.

BROCCOLI

Consumer information   Available fresh all year, with peak October to May. Available frozen all year. Broccoli should be fresh and green, with compact bud clusters that have not opened to show yellow flowers. Overly mature broccoli is lighter in color and is tougher and stronger in taste, having less flavor than the young vegetable. The color can be dark green, sage green, or purple-green, depending on the variety.

Storing   Keep in refrigerator and use as soon as possible.

Uses   Serve raw as an appetizer or in salads. Steam, sauté, or deep-fry in batter. Use in casseroles (baked au gratin.)

Tips   If stems are thick, they can be slit lengthwise halfway up. A few slices of stale bread in the cooking water will minimize the odor.

BRUSSELS SPROUTS

Consumer information   Available fresh all year, with peak September through February. Available all year canned or frozen. Look for sprouts that are firm, compact, and bright green in color. Puffy or soft sprouts are undesirable. Wilted or yellow leaves indicate aging.

Storing   Keep in refrigerator and use as soon as possible.

Uses   Remove tip of stems and pull off loose leaves. Sprouts can be steamed, baked, or sautéed (sliced). They can be served as appetizers, in casseroles, or plain; hot, or cold in salads.

CABBAGES

Consumer information   Available fresh all year. Available all year canned or bottled. Cabbage heads should be fairly solid, heavy, and closely trimmed, with stems cut close to the head. They should have only 3 or 4 outer (wrapper) leaves and preferably no loose leaves (although early cabbage is less solid than later crop). The outer leaves should have a fresh appearance. Avoid cabbage with worm holes or puffiness.

Storing   Keep in refrigerator and use within a week or two at the most.

Uses   Serve raw in salads or coleslaw or juiced; pickled for sauerkraut. Steam, sauté, or bake; stuff. Use whole leaves, cut in large chunks, or shred.

CARROTS

Consumer information   Available fresh, all year, or canned, bottled, or frozen. Young, slender carrots are preferable for salads and vegetable dishes; older ones are better for soups. Carrots should be firm and smooth, and have a rich orange color. Wilted, flabby carrots and those with rough or cracked skins are not desirable.

Storing   Keep in refrigerator and use as required. They will last for several weeks if stored immediately.

Uses   Use raw carrots sliced, grated, or curled (soak in ice water after cutting into thin strips) in salads and as appetizers; or juiced. Steam, sauté, or bake; use whole or pureed in soups. Add to cakes, cookies, breads, soufflés, and custards.

CAULIFLOWERS

Consumer information   Available fresh or frozen all year. Select white cauliflower; a yellowed or speckled surface means it is not fresh. If there are any leaves, they should be fresh and green. Avoid loose, open flower clusters, which are a sign of overmaturity.

Storing   Keep in refrigerator and use as soon as possible.

Uses   Cauliflower can be used whole or broken into flowerets; cooked, or raw in salads. (Also see *Broccoli*.)

CELERIAC

Consumer information   Available fresh all year, with peak October through April. Often called celery root or knob. Celeriac should be firm, clean, and free from blemishes.

Storing   Keep in refrigerator and use as required. It will last for several weeks.

Uses   (see *Celery*.)

CELERY

Consumer information   Available fresh all year. Celery should be fresh, crisp, and free of bruises. Stalks should be thick and solid, with a good heart formation. Excessively hard branches may be stringy and woody, and celery with seed stems is too mature.

Storing   Keep in refrigerator and use as required.

Uses   Use raw in salads and sandwiches, and as an appetizer, or juiced. Steam, sauté, braise, or bake; use in soups.

CELERY ROOT (see Celeriac)

CHARD (see Spinach)

CHICK-PEAS (see Beans, Dried)

CHICORY (see Endives)

CHINESE CABBAGE

Consumer information   Available fresh all year. Heads should be compact, with crisp, clean, fresh green leaves; avoid yellowing or wilted leaves.

Storing   Keep in refrigerator and use within a week.

Uses   Use shredded (raw) in salads; steamed, sautéed or baked.

CHIVES

Consumer information   Available fresh whenever they can be found; usually sold planted in pots. Available frozen all year.

Storing   Keep fresh chives in pots and water daily. Snip off chives as required; with good care, they will last for several weeks.

Uses   In salads, soups, and appetizers. Frozen chives may be used immediately upon removing from freezer and need not be defrosted first.

COLLARDS (see Cabbages)

CORN

Consumer information   Available fresh all year, with peak in spring months. Available canned and frozen all year. Corn can be white or yellow (yellow has more vitamin A). Sweet, not field, corn is more satisfactory and sweeter. If corn has husks, they should be fresh and green, not dry and straw-colored. Avoid buying corn that has been husked.

Storing   Corn should be eaten as soon after picking as possible and should be kept refrigerated continuously.

Uses   On the cob, roast (pull out silk but leave husks on and roast in oven at 325° F. for about 50 minutes), or steam. Steam, sauté, or bake kernels. Use in soups, puddings, fritters, or plain. Raw corn can be ground into meal.

CUCUMBERS

Consumer information Available fresh all year, with peak May through August. Cucumbers should have soft and immature seeds (avoid large ones, which tend to have hard seeds and too much water). Avoid cucumbers that have a yellow color, indicating old age, or show puffiness, or are withered or shriveled.

Storing Keep in refrigerator and use within a few days.

Uses Use raw, sliced or chopped, in salads and sandwiches and as appetizers; add to punches or juices. Score cucumbers from top to bottom with a sharp-tined fork and cut from the middle toward the ends, discarding the ends. Steam and sauté for omelets and pancake fillings. Cucumbers may be pickled.

EGGPLANTS

Consumer information Available fresh all year, with peak August through September. Eggplants should be firm and heavy for their size, with dark-purple or purple and white skin, free of scars or cuts and worm injury. Avoid wilted, shriveled, or soft fruit, which is usually bitter.

Storing Keep in refrigerator and use as soon as possible.

Uses Use whole, cut in slices, chopped, or mashed. Steam, bake, sauté (plain or breaded). Serve au gratin, stuffed, in casseroles, pureed in soufflés; chilled or marinated in salads and appetizers.

Tip To prevent eggplant from discoloring, drop into salted water as soon as it is peeled or rub with lemon juice.

ENDIVES (Also ESCAROLE, CHICORY)

Consumer information Available fresh all year. All greens should be fresh, crisp, cold, and clean and should not have dry or yellowing leaves or seed stems, which indicate old age. Bunches should not show black or otherwise discolored leaf margins or reddish discoloration of the hearts.

Storing Keep in refrigerator and use as soon as possible.

Uses Raw in salads and sandwiches. Steam, braise, bake, sauté, and add to other vegetables.

ESCAROLE (see Endives)

GARLIC

Consumer information Available fresh and dried (powder and salt) all year. Fresh garlic should be firm and white or light purple, depending on variety, without blemishes and with a pungent odor.

Storing Wrap and store fresh garlic in refrigerator, where it will keep for several weeks.

Uses Raw, minced, or pressed in dressings, sauces. Steamed, baked, or sautéed in combination with vegetables, grains, pasta, and soups.

HORSERADISH

Consumer information Available all year. Select firm (not shriveled) roots with no soft spots. Horseradish that has been harvested late, when cold weather has set in, will keep better.

Storing Keep in refrigerator and use as required.

Uses Raw, grated, for sauces, dressings, and soups.

JERUSALEM ARTICHOKES

Consumer information Season is uncertain, and availability fresh may be anytime during the year.

Storing Keep in refrigerator and use as required; they will shrivel if exposed to air for any length of time.

Uses Raw, grated, or finely sliced in salads. Steam, sauté, or bake in the same manner as potatoes.

KALE (see Spinach)

KOHLRABI

Consumer information Available fresh May through November, with peak in June and July. The leaves of kohlrabi are similar to those of the turnip and can be eaten if crisp and green. Stems should be firm and crisp and not too large.

Storing Keep in a cool place and use within a few days.

Uses Steam, sauté, or bake; boil in soups. Serve hot plain, with butter or sauces, or stuffed.

LEEKS

Consumer information Available all year, with peak September through November and in the spring months. Select leeks with fresh, green tops and medium-sized necks, which, if blanched, are whitish green, in length 2 to 3 inches from root. Be sure that leeks "give" to the touch and are not woody inside.

Storing Keep in refrigerator and use within a week.

Uses Wash leeks thoroughly under running water to remove sand, which lodges between the flat leaves. Slice leeks lengthwise and use white and green parts or white parts alone. Leeks may be steamed, braised, sautéed, or baked. They can be a mild substitute for onions in any recipe and are essential in vichyssoise.

LETTUCE

Consumer information Available fresh all year. There are many varieties of lettuce, some available at different times of the year. One of the most com-

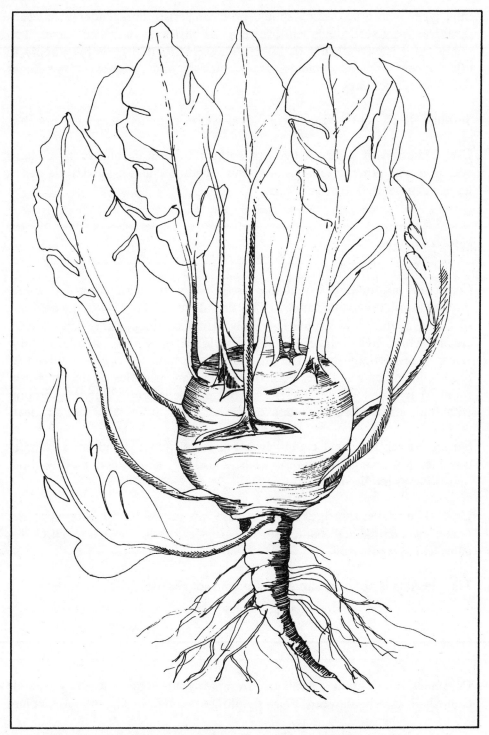

The globe of the kohlrabi is not a root but the enlarged base of the stem.

mon types is iceberg, which should be clean, crisp, and tender with heads firm; free from seed stems and brown areas on leaves or wilted leaves. Too many outer leaves are not desirable. These instructions generally apply to other types, such as romaine, Bibb, Boston, and loose-leaf (sometimes called salad) lettuces.

Storing Keep tightly wrapped in the refrigerator and use as soon as possible.

Uses Use raw in salads, appetizers, and sandwiches. Steam, sauté, or bake and serve with other vegetables or sauces, or stuff. The outside dark leaves of lettuce can be used as soup greens.

MUSHROOMS

Consumer information Available fresh all year, with peak November through April. Available canned or bottled all year. There is usually only one species available; it is white. The flesh should be thick, firm, and white to creamy white, free from open caps, discoloration, wilting, or other injury. If caps are partially open, the gills should be light in color, as brown or black gills indicate old age. Sometimes there are varieties of darker mushrooms that may look a bit shriveled but are definitely edible. Mushrooms with long stems should be less expensive than those with shorter ones.

Storing Keep in a tight container or bag in refrigerator and use as soon as possible. Mushrooms may be frozen, sliced or whole, and used without defrosting as required.

Uses Use whole, sliced, or chopped. Use raw in salads and as an appetizer (plain or marinated). Sauté, steam, deep-fry, broil, or bake; puree for pâtés and spreads; stuff.

Tip Peeling is not necessary, but if preferred, peel only dark mushrooms.

OKRA

Consumer information Available fresh May through October. Available canned or frozen all year. Pods should be young, fresh, and of medium size, 2 to 4 inches long. They should snap or puncture easily, indicating tenderness. Avoid dull-colored, discolored, dried, or shriveled pods.

Storing   Keep in refrigerator and use as soon as possible.

Uses   Steam, sauté, or boil in stews (especially gumbos) and soups. Okra may be sliced and strung up to dry for later use as a condiment.

ONIONS, DRY

Consumer information   Available fresh and dried (powder or salt) all year. Onions should be hard and well shaped, with dry skins that crackle and without seed stems or sprouts. Avoid onions that are thick and woody, or open at the neck, and those that show stem development. Moisture at the neck indicates decay.

Storing   Keep in refrigerator or at room temperature. In any event, onions should be kept dry and not allowed to sprout.

Uses   Use raw in salads (red are particularly good for this), sandwiches, and appetizers; steam, bake, sauté, and deep-fry; boil in soups and stews; stuff. Use whole, sliced, chopped, pureed.

ONIONS, GREEN

Consumer information   Available all year, with peak May through August. A green onion (sometimes called scallion) is an onion harvested very young. There are a number of types, but all should be green, with fresh tops and medium-sized necks, and well blanched for 2 or 3 inches from root and crisp and tender. Wilted or discolored types indicate poor quality.

Storing   Keep in refrigerator and use as soon as possible.

Uses   Use mainly in salads. Other uses are same as for the yellow or red onion.

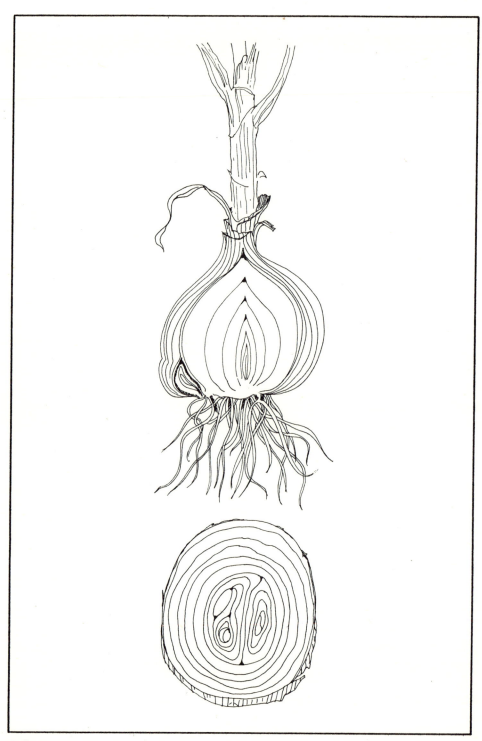

Onions

PARSLEY

Consumer information Available fresh or dried all year. Parsley should be bright, fresh, green, crisp, and free from yellow leaves or dirt. Wilting and yellowing means old age and damage. May be dried by placing in paper bag and hanging by a string. Shake bag once a day.

Storing To dry, see above. Keep fresh parsley in refrigerator and use as desired. There is no advantage in storing with stems in water. Parsley root is sold separately. It keeps best in the refrigerator.

Uses Chop and add to salads, vegetable dishes, soups, and appetizers; also use as a garnish.

PARSNIPS

Consumer information Available fresh mostly October through April. Parsnip flavor is not fully developed until after prolonged exposure to temperature around 40° F. or lower. Parsnips should have been stored, not come direct from the field. They should be smooth, firm, clean, well shaped, and of small to medium size. Softness may be an indication of decay and discoloration may be an indication of freezing.

Storing Keep in refrigerator and use as desired.

Uses Tender and sweet parsnips should be served as a vegetable in their own right rather than a component of stews and soups. Treat them as you do carrots, except that they require shorter cooking. Serve precooked or raw in salads.

PEAS, GREEN

Consumer information Available March through July. Look for young pods, uniformly green and well filled.

Storing Keep in refrigerator and use as soon as possible.

Uses Shell and steam, sauté, bake, or boil. Add to soups, stews, sauces, in pastries, pureed in soups, with other vegetables, raw or cooked in salads. Fresh peas will keep green while cooking if you raise the lid occasionally. Never use baking soda, even though it may keep them green!

PEPPERS, SWEET

Consumer information Available all year, with peak in May through October. Should be fresh, firm, bright, thick-fleshed, and either bright green or red. Immature peppers are usually soft and dull-looking. Bell peppers are the most used, 4 or 5 inches long, with 3 or 4 lobes that taper only slightly toward the blossom end.

Storing Refrigerate immediately and use as quickly as possible.

Uses Use whole, cut in quarters, sliced, or chopped. Can be steamed, sautéed, deep-fried in batter, boiled, baked, or broiled. Stuff and bake them or add to casseroles, stews, and sauces. Serve in combination with other vegetables, pastas, or rice. Serve raw in salads and as an appetizer.

POTATOES

Consumer information Available fresh or frozen all year. Should be firm, relatively smooth, well shaped, not badly cut or bruised or skinned. They should show no green and should not be wilted or show sprouts. Cooking quality depends on variety. Some potatoes tend to be better for baking because of their high content of dry matter. For boiling, slightly higher moisture content is desirable. Do not refrigerate. Low temperature causes conversion of starch to sugar.

Storing Keep away from heat but not too cold, preferably at room temperature, and in the dark.

Uses Use whole, cut in quarters, slice, chop, grate; steam, boil, sauté, fry, deep-fry, bake, roast. Serve with butter or sauces, mashed, and in combination with other vegetables. Add to soups, stews, casseroles, puddings. Add precooked to salads.

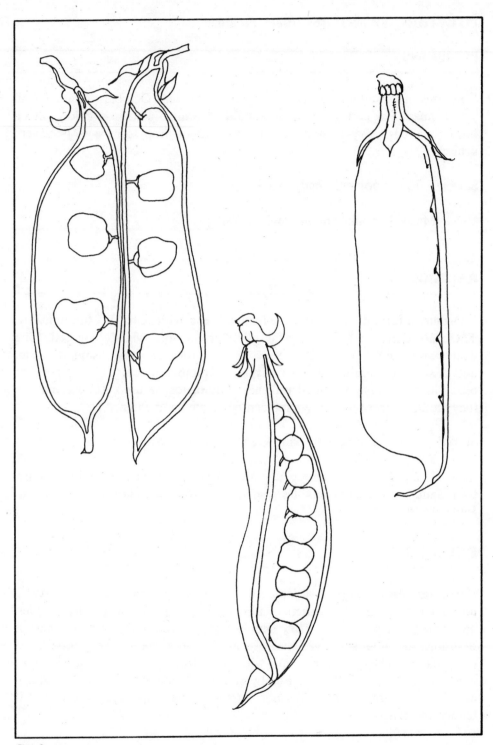
Garden peas

POTATOES, SWEET (see Sweet Potatoes)

PUMPKINS

Consumer information Available fresh mostly late October. Available all year canned. For eating, pumpkin should be orange-golden in color, have a hard rind, be heavy in relation to size, and be free from cuts or severe bruises.

Storing Keep cool until ready to use.

Uses Mashed or baked in pies; boiled in soups.

RADISHES

Consumer information Available fresh all year, with peak May through July. There are different types, such as globular red and white, long red, long white, and long black. All should be well formed, fresh, smooth, firm, and crisp, and should not have any large number of cuts, black spots, or pits. Pithy, spongy, or wilted radishes should *not* be used. If tops are on, they should be fresh and green and can be cooked and eaten.

Storing Remove tops, if any, and keep in refrigerator.

Uses Use whole, sliced, chopped, grated. Serve raw juiced, in salads, as appetizer. Sauté with other vegetables (particularly for oriental dishes).

RHUBARB

Consumer information Available fresh mostly January through June. Available all year frozen or canned. Stalks should be fresh, firm, crisp, tender, and bright in appearance. They should not be excessively thin. Redness is desirable, although some varieties of soil give little red and yet rhubarb has good flavor. The younger stems with immature leaves are usually most tender and delicate in flavor. Flabby stalks indicate stringiness, lack freshness, and are not flavorful. *Do not eat the leaves.* Puncture stalk to tell if tender and crisp.

Storing Keep in refrigerator.

Uses As a dessert in pies, combined with apples, as a stewed fruit.

RUTABAGAS

Consumer information Available fresh in cold months. Rutabagas are mostly yellowish in color, which distinguishes them from the whiter-fleshed turnips, are slightly more elongated than these, and have a thick neck. They should be thick, fresh-looking, and heavy for their size, although size is not a quality factor. Skin must not show any signs of puncture or decay.

Storing Keep in refrigerator and use as desired.

Uses Cooked as a vegetable; also raw as an appetizer.

SHALLOTS

Consumer information Available fresh mainly November through April. Shallots are distinguished from green onions by the bulb made up of cloves like garlic. Good-quality shallots have green, fresh tops and medium-sized necks, which are well blanched for at least 2 or 3 inches from root. Shallots should be young and crisp and tender; wilted and yellowing tops indicate old age.

Storing Keep in refrigerator and use as soon as possible.

Uses Use like onions, raw and cooked in soups and stews or as a vegetable. Shallots' delicate flavor make them more desirable than onions in some dishes.

SPINACH

Consumer information Available fresh all year, with peak in January through May. Available all year canned or frozen. Leaves should be a good green color, clean and fresh-looking, with little or no yellowish green. Larger leaves with yellow discoloration are unacceptable, as are wilted, bruised, or crushed leaves. Plants should be relatively stocky. Straggly, undergrown plants or those with seed stems are undesirable. Most spinach is of the crumpled-leaf or savory kind. Spinach also comes prewashed in cellophane bags.

Storing Because spinach is very perishable, put it in refrigerator immediately and use it as soon as possible. Wash leaves well to remove all sand.

Uses Use whole leaves and stems or only leaves. Tear or chop. Serve raw in salads or juiced. Sauté, steam, deep-fry, boil in soups. Puree in soufflés, puddings, casseroles. Serve with other vegetables, pastas, or rice or in pie fillings.

SQUASH

Consumer information Available fresh all year; large, hard-skinned types such as Hubbard abundant in fall and winter. There are many varieties of squash, usually classified as summer or winter. It is easier to classify types into (1) soft-skinned (immature and small); (2) hard-skinned (mature and small); (3) hard-skinned (mature and large). Soft-skinned squashes (like zucchini), which may be eaten with the skin and pits, should be young, tender, crisp, fresh, and fairly heavy in relation to size; tenderness is the most important factor. The hard-skinned types (such as butternut or acorn) should not have any softness of rind; this indicates immaturity and thin flesh.

Storing Soft-skinned squash should be kept in the refrigerator and used as soon as possible. Hard-skinned types can be kept at room temperature and used as needed, but do not keep them in a warm room.

Uses All squash can be used whole, sliced, or chopped. Hard-skinned are good for baking, plain or stuffed; or grating into pancakes. Soft-skinned can be sautéed quickly, broiled, or steamed. Grate or slice thinly, raw, for salads and appetizers.

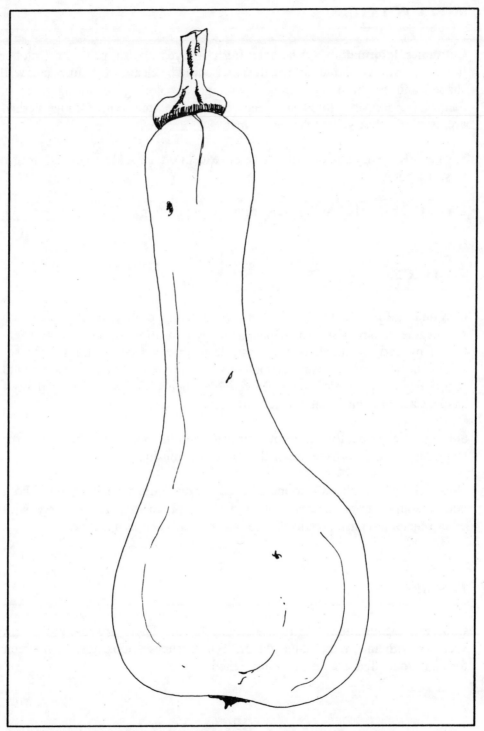

The Hubbard winter squash is large, pale yellow, and sometimes called butternut.

SWEET POTATOES

Consumer information Available fresh all year, with peak in October through April. Available canned all year. Should be clean, fresh, firm, and well shaped and bright in appearance, with good coloring according to type. Some have copper-colored skin and some lighter tan skin. Orange-fleshed are usually softer; yellow-fleshed, firmer.

Storing Keep dry and do not refrigerate except after cooking; cold is harmful to sweet potatoes.

Uses Cooked or baked as potato, or in sweet potato pie.

TOMATOES

Consumer information Available fresh and canned all year. If picked with any degree of pinkness (vine-ripened), they can attain good taste and texture if properly handled. It is impossible to give a formula for tomato selection because of the various methods of shipment. Find a retailer who regularly stocks tomatoes that are reliable, then pick out those of desired shape, size, and color that are not too ripe.

Storing Keep tomatoes at room temperature in the open until they are at the stage desired, then refrigerate or they will turn mushy.

Uses Use whole, sliced, chopped, juiced; steam, sauté, grill, bake, broil, boil; add to soups, stews, sauces; stuff and bake; puree in soufflés. Serve with other vegetables, pasta, or rice. Serve raw in salads and as appetizer.

TURNIPS

Consumer information Available fresh all year. There are many different varieties with much the same flavor. For freshness, look for crisp green tops and roots that are firm and not wilted.

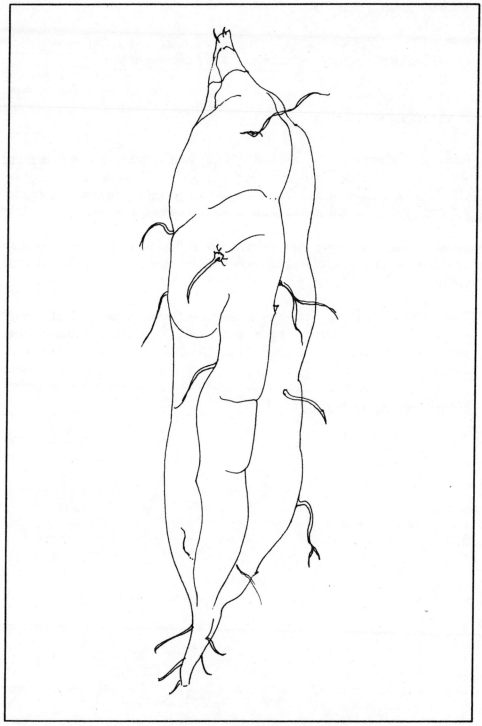

In the U.S., some types of sweet potatoes are erroneously called "yams."

Storing   Keep in refrigerator.

Uses   Use whole, sliced, grated; steam, sauté, bake, or fry. Boil in soups and stews; deep-fry in batter; mash alone or with other vegetables.

WATERCRESS

Consumer information   Available fresh all year in small amounts. Leaves should be fresh, rich green, and free from dirt or yellowishness. Wilting or discoloration denotes old age. There is only one variety of true watercress, but sometimes wild cresses and land cresses can be found in stores.

Storing   Keep in refrigerator. Watercress is highly perishable so use as quickly as possible. Do not soak in water but enclose in plastic wrap or bag with ice.

Uses   Use whole leaves and stems or leaves only. Serve raw in salads, appetizers, sandwiches; or juice. Steam, sauté, boil, bake. Add to soups; serve with other vegetables.

YAMS (see Sweet Potatoes)

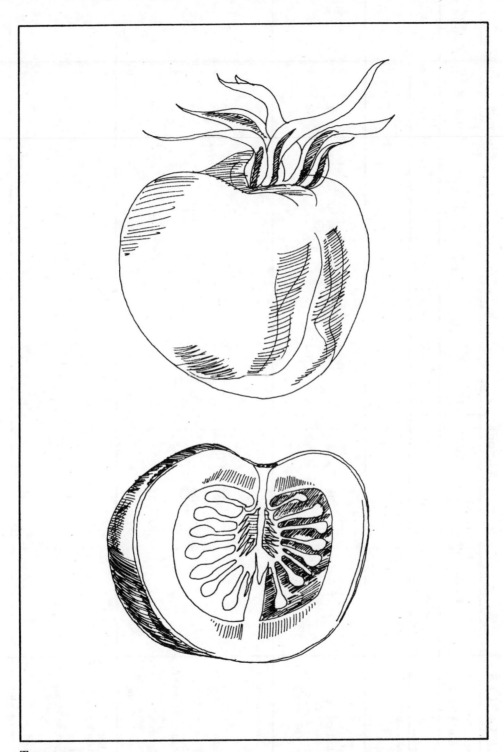

Tomatoes

COMPOSITION OF VEGETABLES

Amounts in 100 grams, edible portion

VEGETABLES FOOD AND DESCRIPTION	VIT. A	Vitamins								Minerals				
		THIAMIN B$_1$	RIBOFLAVIN B$_2$	NIACIN	PYRIDOXINE B$_6$	FOLIC ACID	VIT. C	CALCIUM	PHOSPHORUS	IRON	SODIUM	POTASSIUM	MAGNESIUM
	Internat'l Units	Mg.	Mg.	Mg.	Micro.	Micro.	Mg.	Mg.	Mg.	Mg.	Mg.	Mg.	Mg.
Artichokes, globe or French:													
Raw	160	.08	.05	1.0	—	—	12	51	88	1.3	43	430	—
Cooked	150	.07	.04	.7	—	—	8	51	69	1.1	30	301	—
Asparagus:													
Raw spears	900	.18	.20	1.5	136	109.0	33	22	62	1.0	2	278	20
Cooked spears	900	.16	.18	1.4	—	109.0	26	21	50	.6	1	183	—
Canned spears:													
Green, solids and liquid	510	.06	.09	.8	75	27.0	15	18	43	1.7	236	166	—
White (bleached), solids and liquid	50	.05	.06	.7	—	—	15	15	33	.9	236	140	—
Beans, common, mature seeds, dry:													
White:													
Raw	0	.65	.22	2.4	560	125.0	0	144	425	7.8	19	1,196	170
Cooked	0	.14	.07	.7	—	—	—	50	148	2.7	7	416	—
Red:													
Raw	20	.51	.20	2.3	441	180.0	—	110	406	6.9	10	984	163
Cooked	trace	.11	.06	.7	—	—	—	38	140	2.4	3	340	—
Canned, solids and liquid	trace	.05	.04	.6	—	—	—	29	109	1.8	3	264	—
Pinto, calico and red Mexican, raw	—	.84	.21	2.2	—	—	—	135	457	6.4	10	984	—
Others, including black, brown, and Bayo, raw	30	.55	.20	2.2	—	—	—	135	420	7.9	25	1,038	—
Beans, lima:													
Immature seeds:													
Raw	290	.24	.12	1.4	170	34.0	29	52	142	2.8	2	650	67
Cooked	280	.18	.10	1.3	—	34.0	17	47	121	2.5	1	422	—
Canned, solids and liquid	130	.04	.04	.5	81	13.0	7	26	67	2.4	236	222	—
Mature seeds, dry:													
Raw	trace	.48	.17	1.9	580	103.0	—	72	385	7.8	4	1,529	180
Cooked	—	.13	.06	.7	—	108.0	—	29	154	3.1	2	612	—

VEGETABLES

Amounts in 100 grams, edible portion

FOOD AND DESCRIPTION	VIT. A	Vitamins							Minerals					
		THIAMIN B₁	RIBO-FLAVIN B₂	NIACIN	PYRI-DOXINE B₆	FOLIC ACID	VIT. C	CAL-CIUM	PHOS-PHORUS	IRON	SODIUM	POTAS-SIUM	MAG-NESIUM
	Internat'l Units	Mg.	Mg.	Mg.	Micro.	Micro.	Mg.	Mg.	Mg.	Mg.	Mg.	Mg.	Mg.
Beans, mung:													
Mature seeds, dry, raw	80	.38	.21	2.6	—	—	—	118	340	7.7	6	1,028	—
Sprouted seeds:													
Uncooked	20	.13	.13	.8	—	—	19	19	64	1.3	5	223	—
Cooked	20	.09	.10	.7	—	145.0	6	17	48	.9	4	156	—
Beans, snap (string beans):													
Green:													
Raw	600	.08	.11	.5	63	27.5	19	56	44	.8	7	243	32
Cooked, small amount of water, short time	540	.07	.09	.5	80	28.0	12	50	37	.6	4	151	—
Canned, solids and liquid	290	.03	.04	.3	43	12.0	4	34	21	1.2	236	95	—
Yellow or Wax:													
Raw	250	.08	.11	.5	—	—	20	56	43	.8	7	243	—
Cooked	230	.07	.09	.5	—	32.0	13	50	37	.6	3	151	—
Canned, solids and liquid	60	.03	.04	.3	42	—	5	34	21	1.2	236	95	—
Beets, common red:													
Raw	20	.03	.05	.4	65	14.0	10	16	33	.7	60	335	25
Cooked	20	.03	.04	.3	55	14.0	6	14	23	.5	43	208	—
Canned, solids and liquid	10	.01	.02	.1	54	2.8	3	14	17	.6	236	167	—
Beetgreens, common:													
Raw	6,100	.10	.22	.4	—	60.0	30	119	40	3.3	130	570	106
Cooked	5,100	.07	.15	.3	100	60.0	15	99	25	1.9	76	332	—
Blackeye peas (including cowpeas):													
Immature seeds:													
Raw	370	.43	.13	1.6	118	41.0	29	27	172	2.3	2	541	—
Cooked	350	.30	.11	1.4	95	41.0	17	24	146	2.1	1	379	—
Canned, solids and liquid	60	.09	.05	.5	53	26.0	3	18	112	1.5	236	352	—
Young pods, with seeds:													
Raw	1,600	.15	.14	1.2	—	—	33	65	65	1.0	4	215	—
Cooked	1,400	.09	.09	.8	—	—	17	55	49	.7	3	196	—
Mature seeds, dry:													
Raw	30	1.05	.21	2.2	—	—	—	74	426	5.8	35	1,024	230
Cooked	10	.16	.04	.4	562	44.0	—	17	95	1.3	8	229	—

continued

Amounts in 100 grams, edible portion

VEGETABLES

FOOD AND DESCRIPTION	Vitamins									Minerals						
	VIT. A	THIA-MIN B_1	RIBO-FLAVIN B_2	NIACIN	PYRI-DOXINE B_6	FOLIC ACID	VIT. C	CAL-CIUM	PHOS-PHORUS	IRON	SODIUM	POTAS-SIUM	MAG-NESIUM
	Internat'l Units	Mg.	Mg.	Mg.	Micro.	Micro.	Mg.	Mg.	Mg.	Mg.	Mg.	Mg.	Mg.
Broad beans, raw:													
Immature seeds	220	.28	.17	1.6	—	—	30	27	157	2.2	4	471	—
Mature seeds, raw	70	.50	.30	2.5	—	—	—	102	391	7.1	—	—	—
Broccoli:													
Raw spears	2,500	.10	.23	.9	171	54.0	113	103	78	1.1	15	382	24
Cooked	2,500	.09	.20	.8	—	54.0	90	88	62	.8	10	267	—
Brussels sprouts:													
Raw	550	.10	.16	.9	162	49.0	102	36	80	1.5	14	390	29
Cooked	520	.08	.14	.8	160	49.0	87	32	72	1.1	10	273	—
Cabbage, common:													
Raw	130	.05	.05	.3	120	32.3	47	49	29	.4	20	233	13
Cooked	130	.04	.05	.3	120	32.3	33	44	20	.3	14	163	—
Red, raw	40	.09	.06	.4	—	—	61	42	35	.8	26	268	—
Savoy, raw	200	.05	.08	.3	—	—	55	67	54	.9	22	269	—
Cabbage, Chinese (celery cabbage or petsai), raw	150	.05	.04	.6	—	—	25	43	40	.6	23	253	14
Cabbage, Spoon or Pakchoy:													
Raw	3,100	.05	.10	.8	—	—	25	165	44	.8	26	306	—
Cooked	3,100	.04	.08	.7	—	—	15	148	33	.6	18	214	—
Carrots:													
Raw	11,000	.06	.05	.6	120	8.0	8	37	36	.7	47	341	23
Cooked	10,000	.05	.05	.5	—	—	6	33	31	.6	33	222	—
Canned, solids and liquid	10,000	.02	.02	.4	41	3.3	2	25	20	.7	236	120	—
Cauliflower:													
Raw	60	.11	.10	.7	177	22.2	78	25	56	1.1	13	295	24
Cooked	60	.09	.08	.6	—	—	55	21	42	.7	9	206	—
Celeriac root, raw	—	.05	.06	.7	—	—	8	43	115	.6	100	300	—

Amounts in 100 grams, edible portion

VEGETABLES

FOOD AND DESCRIPTION	VIT. A	Vitamins						Minerals					
		THIAMIN B_1	RIBOFLAVIN B_2	NIACIN	PYRIDOXINE B_6	FOLIC ACID	VIT. C	CALCIUM	PHOSPHORUS	IRON	SODIUM	POTASSIUM	MAGNESIUM
	Internat'l Units	Mg.	Mg.	Mg.	Micro.	Micro.	Mg.	Mg.	Mg.	Mg.	Mg.	Mg.	Mg.
Celery, including green and yellow varieties:													
Raw	240	.03	.03	.3	60	7.0	9	39	28	.3	126	341	22
Cooked	230	.02	.03	.3	—	—	6	31	22	.2	88	239	—
Chard, Swiss:													
Raw	6,500	.06	.17	.5	—	50.0	32	88	39	3.2	147	550	65
Cooked	5,400	.04	.11	.4	—	42.0	16	73	24	1.8	86	321	—
Chervil, raw	—	—	—	—	—	—	9	—	—	—	—	—	—
Chickpeas or garbanzos, mature seeds, dry, raw	50	.31	.15	2.0	—	—	—	150	331	6.9	26	797	—
Chicory, Witloof (French or Belgian endive), bleached head, raw	trace	—	—	—	—	—	—	18	21	.5	7	182	13
Chicory greens, raw	4,000	.06	.10	.5	45	28.0	22	86	40	.9	—	420	—
Chives, raw	5,800	.08	.13	.5	—	—	56	69	44	1.7	—	250	32
Collards: Raw:													
Leaves, without stems	9,300	.16	.31	1.7	—	—	152	250	82	1.5	—	450	57
Leaves, including stems	6,500	.20	.31	1.7	—	—	92	203	63	1.0	43	401	—
Cooked, in small amount of water:													
Leaves, without stems	7,800	.11	.20	1.2	195	102.0	76	188	52	.8	—	262	—
Leaves, including stems	5,400	.14	.20	1.2	—	—	46	152	39	.6	25	234	—
Corn, sweet:													
Raw, white and yellow	400	.15	.12	1.7	222	28.0	12	3	111	.7	trace	280	48
Cooked, kernels cut off cob before cooking	400	.11	.10	1.3	161	28.0	7	3	89	.6	trace	165	—
Kernels cooked on cob	400	.12	.10	1.4	161	28.0	9	3	89	.6	trace	196	—
Canned, solids and liquid	330	.03	.05	1.0	200	8.0	5	3	73	.5	236	97	—

continued

Amounts in 100 grams, edible portion

VEGETABLES		Vitamins							Minerals					
FOOD AND DESCRIPTION	VIT. A	THIAMIN B_1	RIBO-FLAVIN B_2	NIACIN	PYRIDOXINE B_6	FOLIC ACID	VIT. C	CALCIUM	PHOSPHORUS	IRON	SODIUM	POTASSIUM	MAGNESIUM
	Internat'l Units	Mg.	Mg.	Mg.	Micro.	Micro.	Mg.	Mg.	Mg.	Mg.	Mg.	Mg.	Mg.
Cress, garden:													
Raw	9,300	.08	.26	1.0	—	—	69	81	76	1.3	14	606	—
Cooked, small amount of water, short time	7,700	.06	.16	.8	—	—	34	61	48	.8	8	353	—
Cucumber, raw:													
Not pared	250	.03	.04	.2	42	7.0	11	25	27	1.1	6	160	11
Pared	trace	.03	.04	.2	—	—	11	17	18	.3	6	160	—
Dandelion greens:													
Raw	14,000	.19	.26	—	—	—	35	187	66	3.1	76	397	36
Cooked	11,000	.13	.16	—	—	—	18	140	42	1.8	44	232	—
Eggplant:													
Raw	10	.05	.05	.6	—	10.0	5	12	26	.7	2	214	16
Cooked	10	.05	.04	.5	81	10.0	3	11	21	.6	1	150	—
Endive (curly and escarole), raw	3,000	.07	.14	.5	20	47.0	10	81	54	1.7	14	294	10
Fennel, common, leaves, raw	3,500	—	—	—	—	—	31	100	51	2.7	—	397	—
Garlic, cloves, raw	trace	.25	.08	.5	—	—	15	29	202	1.5	19	529	36
Ginger root, fresh	10	.02	.04	.7	—	—	4	23	36	2.1	6	264	—
Horseradish:													
Raw	—	.07	—	—	—	—	81	140	64	1.4	8	564	34
Prepared	—	—	—	—	—	—	—	61	32	.9	96	290	—
Jerusalem artichoke, raw	20	.20	.06	1.3	—	—	4	14	78	3.4	—	—	11
Kale:													
Raw:													
Leaves, without stems (midribs)	10,000	.16	.26	2.1	185	70.0	186	249	93	2.7	75	378	37

Amounts in 100 grams, edible portion

VEGETABLES

FOOD AND DESCRIPTION	Vitamins								Minerals				
	VIT. A	THIAMIN B₁	RIBOFLAVIN B₂	NIACIN	PYRIDOXINE B₆	FOLIC ACID	VIT. C	CALCIUM	PHOSPHORUS	IRON	SODIUM	POTASSIUM	MAGNESIUM
	Internat'l Units	Mg.	Mg.	Mg.	Micro.	Micro.	Mg.	Mg.	Mg.	Mg.	Mg.	Mg.	Mg.
Leaves, including stems Cooked:	8,900	—	—	—	—	—	125	179	73	2.2	75	378	—
Leaves, without stems (midribs)	8,300	.10	.18	1.6	185	70.0	93	187	58	1.6	43	221	—
Leaves, including stems	7,400	—	—	—	—	—	62	134	46	1.2	43	221	—
Kohlrabi, bulb:													
Raw	20	.06	.04	.3	—	10.0	66	41	51	.5	8	372	37
Cooked	20	.06	.03	.2	—	—	43	33	41	.3	6	260	—
Leeks, bulb and lower leaf portion, raw	40	.11	.06	.5	—	—	17	52	50	1.1	5	347	23
Lentils, mature seeds, dry:													
Whole: Raw	60	.37	.22	2.0	—	99.0	—	79	377	6.8	30	790	80
Cooked	20	.07	.06	.6	—	—	—	25	119	2.1	—	249	—
Split, without seed coat, raw	60	.37	.22	2.0	—	—	—	46	260	6.8	—	—	—
Lettuce, raw:													
Butterhead varieties, as Boston types and Bibb	970	.06	.06	.3	71	21.0	8	35	26	2.0	9	264	—
Cos or romaine, as Dark Green and White Paris	1,900	.05	.08	.4	—	—	18	68	25	1.4	9	264	—
Crisphead varieties, as Iceberg, New York and Great Lakes strains	330	.06	.06	.3	55	21.0	6	20	22	.5	9	175	11
Loose leaf or bunching varieties, as Grand Rapids, Salad Bowl, Simpson	1,900	.05	.08	.4	—	31.0	18	68	25	1.4	9	264	—

continued

Amounts in 100 grams, edible portion

VEGETABLES

FOOD AND DESCRIPTION	VIT. A	Vitamins						Minerals					
		THIA-MIN B_1	RIBO-FLAVIN B_2	NIACIN	PYRI-DOXINE B_6	FOLIC ACID	VIT. C	CAL-CIUM	PHOS-PHORUS	IRON	SODIUM	POTAS-SIUM	MAG-NESIUM
	Internat'l Units	Mg.	Mg.	Mg.	Micro.	Micro.	Mg.	Mg.	Mg.	Mg.	Mg.	Mg.	Mg.
Mushrooms: *Agaricus campestris,* cultivated commercially:													
Raw	trace	.10	.46	4.2	45	24.0	3	6	116	.8	15	414	13
Canned, solids and liquid	trace	.02	.25	2.0	60	4.0	2	6	68	.5	400	197	8
Other edible species, raw	trace	.10	.33	6.8	63	3.5	3	13	97	1.4	10	375	—
Mustard greens:													
Raw	7,000	.11	.22	.8	133	60.0	97	183	50	3.0	32	377	27
Cooked	5,800	.08	.14	.6	133	60.0	48	138	32	1.8	18	220	—
New Zealand spinach:													
Raw	4,300	.04	.17	.6	198	75.0	30	58	46	2.6	159	795	40
Cooked	3,600	.03	.10	.5	130	75.0	14	48	28	1.5	92	463	—
Okra:													
Raw	520	.17	.21	1.0	—	—	31	92	51	.6	3	249	41
Cooked	490	.13	.18	.9	45	24.0	20	92	41	.5	2	174	—
Onions, mature (dry):													
Raw	40	.03	.04	.2	130	10.0	10	27	36	.5	10	157	12
Cooked	40	.03	.03	.2	—	10.0	7	24	29	.4	7	110	—
Onions, young green (bunching varieties), raw:													
Bulb and entire top	2,000	.05	.05	.4	—	14.0	32	51	39	1.0	5	231	—
Bulb and white portion of top	trace	.05	.04	.4	—	—	25	40	39	.6	5	231	—
Tops only (green portion)	4,000	.07	.10	.6	—	—	51	56	39	2.2	5	231	—
Onions, Welsh, raw	—	.05	.09	.4	—	—	27	18	49	—	—	—	—
Parsley, common garden and curled-leaf varieties, raw	8,500	.12	.26	1.2	164	38.0	172	203	63	6.2	45	727	41

Amounts in 100 grams, edible portion

VEGETABLES

FOOD AND DESCRIPTION	Vitamins								Minerals					
	VIT. A	THIAMIN B_1	RIBOFLAVIN B_2	NIACIN	PYRIDOXINE B_6	FOLIC ACID	VIT. C	CALCIUM	PHOSPHORUS	IRON	SODIUM	POTASSIUM	MAGNESIUM
	Internat'l Units	Mg.	Mg.	Mg.	Micro.	Micro.	Mg.	Mg.	Mg.	Mg.	Mg.	Mg.	Mg.
Parsnips:													
Raw	30	.08	.09	.2	—	—	16	50	77	.7	12	541	32
Cooked	30	.07	.08	.1	90	23.0	10	45	62	.6	8	379	—
Peas, edible-podded:													
Raw	680	.28	.12	—	100	—	21	62	90	.7	—	170	—
Cooked	610	.22	.11	—	—	20.0	14	56	76	.5	—	119	—
Peas, green, immature:													
Raw	640	.35	.14	2.9	150	25.0	27	26	116	1.9	2	316	35
Cooked	540	.28	.11	2.3	—	25.0	20	23	99	1.8	1	196	—
Canned, solids and liquid Alaska (Early or June peas)	450	.09	.05	.9	50	10.0	9	20	66	1.7	236	96	—
Sweet (sweet wrinkled peas, sugar peas)	450	.11	.06	1.0	44	10.0	9	19	58	1.5	236	96	—
Peas, mature seeds, dry:													
Whole, raw	120	.74	.29	3.0	—	—	—	64	340	5.1	35	1,500	180
Split, without seed coat:													
Raw	120	.74	.29	3.0	130	—	—	33	268	5.1	40	895	—
Cooked	40	.15	.09	.9	20	51.0	—	11	89	1.7	13	296	—
Peppers, hot chili:													
Immature green, raw pods, excluding seeds	770	.09	.06	1.7	—	—	235	10	25	.7	—	—	—
Canned pods, excluding seeds, solids and liquid	610	.02	.05	.8	—	—	68	7	17	.5	—	—	—
Mature red:													
Raw:													
Pods, including seeds	21,600	.22	.36	4.4	—	—	369	29	78	1.2	—	—	—
Pods, excluding seeds	21,600	.1	.2	2.9	—	—	369	16	49	1.4	25	564	—
Dried pods	77,000	.23	1.33	10.5	—	—	12	130	240	7.8	373	1,201	—

continued

207

Amounts in 100 grams, edible portion

VEGETABLES

FOOD AND DESCRIPTION	Vitamins								Minerals					
	VIT. A	THIA-MIN B_1	RIBO-FLAVIN B_2	NIACIN	PYRI-DOXINE B_6	FOLIC ACID	VIT. C	CAL-CIUM	PHOS-PHORUS	IRON	SODIUM	POTAS-SIUM	MAG-NESIUM
	Internat'l Units	Mg.	Mg.	Mg.	Micro.	Micro.	Mg.	Mg.	Mg.	Mg.	Mg.	Mg.	Mg.
Peppers, sweet, garden varieties:													
Immature green:													
Raw	420	.08	.08	.5	260	7.0	128	9	22	.7	13	213	18
Cooked	420	.06	.07	.5	—	—	96	9	16	.5	9	149	—
Mature, red, raw	4,450	.08	.08	.5	—	—	204	13	30	.6	—	—	—
Pickles, cucumber:													
Dill	100	trace	.02	trace	7	—	6	26	21	1.0	1,428	200	12
Sour	100	trace	.02	trace	—	—	7	17	15	3.2	1,353	—	—
Sweet	90	trace	.02	trace	—	—	6	12	16	1.2	—	—	1
Pigeon peas, raw:													
Immature seeds	140	.40	.17	2.2	—	—	39	42	127	1.6	5	552	—
Mature seeds	80	.32	.16	3.0	—	—	—	107	316	8.0	26	981	121
Pimientos, canned, solids and liquid	2,300	.02	.06	.4	—	—	95	7	17	1.5	—	—	—
Potatoes:													
Raw	trace	.10	.04	1.5	220	6.8	20	7	53	.6	3	407	34
Cooked:													
Baked in skin	trace	.10	.04	1.7	233	—	20	9	65	.7	4	503	—
Boiled in skin	trace	.09	.04	1.5	174	7.0	16	7	53	.6	3	407	—
Canned, solids and liquid	trace	.04	.02	.6	—	—	13	4	30	.3	1	250	—
Pumpkin:													
Raw	1,600	.05	.11	.6	—	10.0	9	21	44	.8	1	340	12
Canned	6,400	.03	.05	.6	56	8.0	5	25	26	.4	2	340	—
Pumpkin and squash seed kernels, dry	70	.24	.19	2.4	—	—	—	51	1,444	11.2	—	—	—
Radish, raw:													
Common	10	.03	.03	.3	75	7.0	26	30	31	1.0	18	322	15
Oriental, including daikon	10	.03	.02	.4	—	—	32	35	26	.6	—	180	—

Amounts in 100 grams, edible portion

VEGETABLES

FOOD AND DESCRIPTION	Vitamins								Minerals					
	VIT. A	THIAMIN B_1	RIBO-FLAVIN B_2	NIACIN	PYRI-DOXINE B_6	FOLIC ACID	VIT. C	CAL-CIUM	PHOS-PHORUS	IRON	SODIUM	POTAS-SIUM	MAG-NESIUM
	Internat'l Units	Mg.	Mg.	Mg.	Micro.	Micro.	Mg.	Mg.	Mg.	Mg.	Mg.	Mg.	Mg.
Rhubarb:													
Raw	100	.03	.07	.3	—	4.0	9	96	18	.8	2	251	16
Cooked	80	.02	.05	.3	25	4.0	6	78	15	.6	2	203	—
Rutabagas:													
Raw	580	.07	.07	1.1	—	5.0	43	66	39	.4	5	239	15
Cooked	550	.06	.06	.8	100	5.0	26	59	31	.3	4	167	—
Salsify:													
Raw	10	.04	.04	.3	—	—	11	47	66	1.5	—	380	—
Cooked	10	.03	.04	.2	—	—	7	42	53	1.3	—	266	—
Sauerkraut, canned, solids and liquid	50	.03	.04	.2	130	—	14	36	18	.5	747	140	—
Seaweed, raw:													
Agar	—	—	—	—	—	—	—	567	22	6.3	—	8,060	—
Dulse	—	—	—	—	—	—	—	296	267	—	2,085	—	—
Irish moss	—	—	—	—	—	—	—	885	157	8.9	2,892	2,844	—
Kelp	—	—	—	—	—	—	—	1,093	240	—	3,007	5,273	—
Laver	—	—	—	—	—	—	—	—	—	—	—	—	—
Shallots, bulbs, raw	trace	.06	.02	.2	—	—	8	37	60	1.2	12	334	—
Soybeans:													
Immature seeds:													
Raw	690	.44	.16	1.4	—	—	29	67	225	2.8	—	—	—
Cooked	660	.31	.13	1.2	—	—	17	60	191	2.5	—	—	—
Canned, solids and liquid	—	.09	.09	—	—	—	8	55	100	2.9	236	—	—
Mature seeds, dry:													
Raw	80	1.10	.31	2.2	—	—	—	226	554	8.4	5	1,677	265
Cooked	30	.21	.09	.6	—	—	0	73	179	2.7	2	540	—
Fermented products:													
Natto (soybeans)	0	.07	.50	1.1	—	—	0	103	182	3.7	—	249	—
Miso (soybeans and cereal)	40	.06	.10	.3	—	—	0	68	309	1.7	2,950	334	—
Sprouted seeds:													
Raw	80	.23	.20	.8	—	—	13	48	67	1.0	—	—	—
Cooked	80	.16	.15	.7	—	—	4	43	50	.7	—	—	—

continued

Amounts in 100 grams, edible portion

VEGETABLES

FOOD AND DESCRIPTION	Vitamins								Minerals					
	VIT. A	THIA-MIN B_1	RIBO-FLAVIN B_2	NIACIN	PYRI-DOXINE B_6	FOLIC ACID	VIT. C	CAL-CIUM	PHOS-PHORUS	IRON	SODIUM	POTAS-SIUM	MAG-NESIUM
	Internat'l Units	Mg.	Mg.	Mg.	Micro.	Micro.	Mg.	Mg.	Mg.	Mg.	Mg.	Mg.	Mg.
Soybean curd (tofu)	0	.06	.03	.1	—	—	0	128	126	1.9	7	42	111
Soy sauce, Tamari	0	.02	.25	.4	—	—	0	82	104	4.8	7,325	366	—
Spinach:													
Raw	8,100	.10	.20	.6	198	75.0	58	93	51	3.1	71	470	88
Cooked	8,100	.07	.14	.5	130	75.0	21	93	38	2.2	50	324	—
Canned, solids and liquid	5,500	.02	.10	.3	95	49.0	14	85	26	2.1	236	250	63
Squash:													
Summer, all varieties:													
Raw	410	.05	.09	1.0	63	17.0	22	28	29	.4	1	202	—
Cooked	390	.05	.08	.8	63	11.0	10	25	25	.4	1	141	—
Zucchini and Cocozelle, green:													
Raw	320	.05	.09	1.0	—	11.0	19	28	29	.4	1	202	—
Cooked	300	.05	.08	.8	—	—	9	25	25	.4	1	141	—
Winter, all varieties:													
Raw	3,700	.05	.11	.6	91	12.0	13	22	38	.6	1	369	17
Cooked:													
Baked	4,200	.05	.13	.7	91	12.0	13	28	48	.8	1	461	—
Boiled, mashed	3,500	.04	.10	.4	—	—	8	20	32	.5	1	258	—
Butternut:													
Raw	5,700	.05	.11	.6	—	—	9	32	58	.8	1	487	—
Cooked:													
Baked	6,400	.05	.13	.7	—	—	8	40	72	1.0	1	609	—
Boiled, mashed	5,400	.04	.10	.4	—	—	5	29	49	.7	1	341	—
Sweet potatoes:													
Raw:													
Firm-fleshed (Jersey types)	9,200	.10	.06	.6	320	12.0	23	32	47	.7	10	243	31
Soft-fleshed (mainly Porto-Rico variety)	8,700	.10	.06	.6	320	12.0	20	32	47	.7	10	243	31
Cooked:													
Baked in skin	8,100	.09	.07	.7	218	12.0	22	40	58	.9	12	300	—
Boiled in skin	7,900	.09	.06	.6	—	—	17	32	47	.7	10	243	—
Canned, liquids and solid, without added sugar or salt	5,000	.03	.03	.6	—	—	8	13	29	.7	12	120	—

Amounts in 100 grams, edible portion

VEGETABLES

FOOD AND DESCRIPTION	Vitamins								Minerals					
	VIT. A	THIAMIN B₁	RIBOFLAVIN B₂	NIACIN	PYRIDOXINE B₆	FOLIC ACID	VIT. C	CALCIUM	PHOSPHORUS	IRON	SODIUM	POTASSIUM	MAGNESIUM
	Internat'l Units	Mg.	Mg.	Mg.	Micro.	Micro.	Mg.	Mg.	Mg.	Mg.	Mg.	Mg.	Mg.
Tomatoes, green, raw	270	.06	.04	.5	—	—	20	—	—	—	—	—	—
Tomatoes, ripe:													
Raw	900	.06	.04	.7	—	8.0	23	13	27	.5	3	244	14
Cooked, boiled	1,000	.07	.05	.8	—	—	24	13	27	.5	3	244	—
Canned, solids and liquid	900	.05	.03	.7	151	3.7	17	15	32	.6	4	287	12
Turnips:													
Raw	trace	.04	.07	.6	98	5.0	36	39	30	.5	49	268	20
Cooked	trace	.04	.05	.3	90	4.0	22	35	24	.4	34	188	—
Turnip greens, leaves, including stems:													
Raw	7,600	.21	.39	.8	—	83.0	139	246	58	1.8	—	—	58
Cooked in small amount of water, short time	6,300	.15	.24	.6	—	—	69	184	37	1.1	—	—	—
Watercress, leaves, including stems, raw	4,900	.08	.16	.9	—	48.0	79	151	54	1.7	52	282	20
Yam, tuber, raw	trace	.10	.04	.5	—	—	9	20	69	.6	—	600	—

COMPOSITION OF VEGETABLES (continued)

VEGETABLES

Amounts in 100 grams, edible portion

FOOD AND DESCRIPTION	WATER	FOOD ENERGY	PRO-TEIN	CARBOHYDRATE TOTAL	CARBOHYDRATE FIBER	ASH	TOTAL FAT	MISCELLANEOUS AND COMMENTS
	percent	calories	grams	grams	grams	grams	grams	
Artichokes, globe or French:								
Raw	85.5	9–47	2.9	10.6	2.4	0.8	0.2	
Cooked	86.5	8–44	2.8	9.9	2.4	.6	.2	
Asparagus:								
Raw spears	91.7	26	2.5	5.0	.7	.6	.2	High in folic acid
Cooked spears	93.6	20	2.2	3.6	.7	.4	.2	
Canned spears:								
Green, solids and liquid	93.6	18	1.9	2.9	.5	1.3	.3	
White (bleached), solids and liquid	93.3	18	1.6	3.3	.5	1.5	.3	
Beans, common, mature seeds, dry:								
White:								
Raw	10.9	340	22.3	61.3	4.3	3.9	1.6	*
Cooked	69.0	118	7.8	21.2	1.5	1.4	.6	
Red:								
Raw	10.4	343	22.5	61.9	4.2	3.7	1.5	
Cooked	69.0	118	7.8	21.4	1.5	1.3	.5	
Canned, solids and liquid	76.0	90	5.7	16.4	.9	1.5	.4	
Pinto, calico, and red Mexican, raw	8.3	349	22.9	63.7	4.3	3.9	1.2	
Others, including black, brown, and Bayo, raw	11.2	339	22.3	61.2	4.4	3.8	1.5	
Beans, lima:								
Immature seeds:								
Raw	67.5	123	8.4	22.1	1.8	1.5	.5	*
Cooked	71.1	111	7.6	19.8	1.8	1.0	.5	
Canned, solids and liquid	80.8	71	4.1	13.4	1.3	1.4	.3	
Mature seeds, dry:								
Raw	10.3	345	20.4	64.0	4.3	3.7	1.6	
Cooked	64.1	138	8.2	25.6	1.7	1.5	.6	

Amounts in 100 grams, edible portion

VEGETABLES

FOOD AND DESCRIPTION	WATER	FOOD ENERGY	PRO-TEIN	CARBOHYDRATE TOTAL	FIBER	ASH	TOTAL FAT	MISCELLANEOUS AND COMMENTS
	percent	calories	grams	grams	grams	grams	grams	
Beans, mung:								
Mature seeds, dry, raw	10.7	340	24.2	60.3	4.4	3.5	1.3	Sprouted bean contains enzymes to aid in digestion.
Sprouted seeds:								
Uncooked	88.8	35	3.8	6.6	.7	.6	.2	
Cooked	91.0	28	3.2	5.2	.7	.4	.2	
Beans, snap (string beans):								
Green:								
Raw	90.1	32	1.9	7.1	1.0	.7	.2	
Cooked, small amount of water, short time	92.4	25	1.6	5.4	1.0	.4	.2	
Canned, solids and liquid	93.5	18	1.0	4.2	.6	1.2	.1	
Yellow or Wax:								
Raw	91.4	27	1.7	6.0	1.0	.7	.2	
Cooked	93.4	22	1.4	4.6	1.0	.4	.2	
Canned, solids and liquid	93.7	19	1.0	4.2	.6	.9	.2	
Beets, common red:								
Raw	87.3	43	1.6	9.9	.8	1.1	.1	High in zinc.
Cooked	90.9	32	1.1	7.2	.8	.7	.1	
Canned, solids and liquid	90.3	34	.9	7.9	.5	.8	.1	
Beet greens, common:								
Raw	90.9	24	2.2	4.6	1.3	2.0	.3	Contains large amounts of oxalic acid.**
Cooked	93.6	18	1.7	3.3	1.1	1.2	.2	
Blackeye peas (including cowpeas):								
Immature seeds:								*
Raw	66.8	127	9.0	21.8	1.8	1.6	.8	
Cooked	71.8	108	8.1	18.1	1.8	1.2	.8	
Canned, solids and liquid	81.0	70	5.0	12.4	.7	1.3	.3	
Young pods, with seeds:								
Raw	86.0	44	3.3	9.5	1.7	.9	.3	
Cooked	89.5	34	2.6	7.0	1.7	.6	.3	
Mature seeds, dry:								
Raw	10.5	343	22.8	61.7	4.4	3.5	1.5	
Cooked	80.0	76	5.1	13.8	1.0	.8	.3	

continued

Amounts in 100 grams, edible portion

VEGETABLES

FOOD AND DESCRIPTION	WATER	FOOD ENERGY	PRO-TEIN	CARBOHYDRATE TOTAL	CARBOHYDRATE FIBER	ASH	TOTAL FAT	MISCELLANEOUS AND COMMENTS
	percent	calories	grams	grams	grams	grams	grams	
Broad beans, raw:								
Immature seeds	72.3	105	8.4	17.8	2.2	1.1	.4	*
Mature seeds, raw	11.9	338	25.1	58.2	6.7	3.1	1.7	
Broccoli:								
Raw spears	89.1	32	3.6	5.9	1.5	1.1	.3	
Cooked	91.3	26	3.1	4.5	1.5	.8	.3	
Brussels sprouts:								
Raw	85.2	45	4.9	8.3	1.6	1.2	.4	
Cooked	88.2	36	4.2	6.4	1.6	.8	.4	
Cabbage, common:								Raw cabbage high in Vitamin C.
Raw	92.4	24	1.3	5.4	.8	.7	.2	
Cooked	93.9	20	1.1	4.3	.8	.5	.2	
Red, raw	90.2	31	2.0	6.9	1.0	.7	.2	
Savoy, raw	92.0	24	2.4	4.6	.8	.8	.2	
Cabbage, Chinese (celery cabbage or petsai) raw	95.0	14	1.2	3.0	.6	.7	.1	
Cabbage, Spoon or Pakchoy:								
Raw	94.3	16	1.6	2.9	.6	1.0	.2	
Cooked	95.2	14	1.4	2.4	.6	.8	.2	
Carrots:								High in sodium and Vitamin A.
Raw	88.2	42	1.1	9.7	1.0	.8	.2	
Cooked	91.2	31	.9	7.1	1.0	.6	.2	
Canned, solids and liquid	91.8	28	.6	6.5	.6	.9	.2	
Cauliflower:								
Raw	91.0	27	2.7	5.2	1.0	.9	.2	
Cooked	92.8	22	2.3	4.1	1.0	.6	.2	
Celeriac root, raw	88.4	40	1.8	8.5	1.3	1.0	.3	

Amounts in 100 grams, edible portion

VEGETABLES

FOOD AND DESCRIPTION	WATER	FOOD ENERGY	PRO-TEIN	CARBOHYDRATE TOTAL	FIBER	ASH	TOTAL FAT	MISCELLANEOUS AND COMMENTS
	percent	calories	grams	grams	grams	grams	grams	
Celery, including green and yellow varieties:								
Raw	94.1	17	.9	3.9	.6	1.0	.1	
Cooked	95.3	14	.8	3.1	.6	.7	.1	
Chard, Swiss:								High in potassium.**
Raw	91.1	25	2.4	4.6	0.8	1.6	0.3	
Cooked	93.7	18	1.8	3.3	.7	1.0	.2	
Chervil, raw	80.7	57	3.4	11.5	—	3.5	.9	
Chickpeas or garbanzos, mature seeds, dry, raw	10.7	360	20.5	61.0	5.0	3.0	4.8	*
Chicory, Witloof (French or Belgian endive), bleached head, raw	95.1	15	1.0	3.2	—	.6	.1	
Chicory greens, raw	92.8	20	1.8	3.8	.8	1.3	.3	
Chives, raw	91.3	28	1.8	5.8	1.1	.8	.3	
Collards:								High in calcium.**
Raw:								
Leaves, without stems	85.3	45	4.8	7.5	1.2	1.6	0.8	
Leaves, including stems	86.9	40	3.6	7.2	.9	1.6	.7	
Cooked, in small amount of water:								
Leaves, without stems	89.6	33	3.6	5.1	1.0	1.0	.7	
Leaves, including stems	90.8	29	2.7	4.9	.8	1.0	.6	
Corn, sweet:								Along with potatoes, corn is highest in starch, except for beans.
Raw, white and yellow	72.7	96	3.5	22.1	.7	.7	1.0	
Cooked, kernels cut off cob before cooking	76.5	83	3.2	18.8	.7	.5	1.0	
Kernels cooked on cob	74.1	91	3.3	21.0	.7	.6	1.0	
Canned, solids and liquid	75.5	83	2.5	20.5	.8	1.0	.5	

continued

Amounts in 100 grams, edible portion
VEGETABLES

FOOD AND DESCRIPTION	WATER	FOOD ENERGY	PRO-TEIN	CARBOHYDRATE TOTAL	CARBOHYDRATE FIBER	ASH	TOTAL FAT	MISCELLANEOUS AND COMMENTS
	percent	calories	grams	grams	grams	grams	grams	
Cress, garden:								
Raw	89.4	32	2.6	5.5	1.1	1.8	.7	**
Cooked, small amount of water, short time	92.5	23	1.9	3.8	.9	1.2	.6	
Cucumber, raw:								
Not pared	95.1	15	.9	3.4	.6	.5	.1	Contains more water than any other raw vegetable.
Pared	95.7	14	.6	3.2	.3	.4	.1	
Dandelion greens:								
Raw	85.6	45	2.7	9.2	1.6	1.8	.7	High in calcium.**
Cooked	89.8	33	2.0	6.4	1.3	1.2	.6	
Eggplant:								
Raw	92.4	25	1.2	5.6	.9	.6	.2	
Cooked	94.3	19	1.0	4.1	.9	.4	.2	
Endive (curly and escarole), raw	93.1	20	1.7	4.1	.9	1.0	.1	
Fennel, common, leaves, raw	90.1	28	2.8	5.1	.5	1.7	.4	
Garlic, cloves, raw	61.3	137	6.2	30.8	1.5	1.5	.2	
Ginger root, fresh	87.0	49	1.4	9.5	1.1	1.1	1.0	
Horseradish:								
Raw	74.6	87	3.2	19.7	2.4	2.2	.3	
Prepared	87.1	38	1.3	9.6	.9	1.8	.2	
Jerusalem artichoke, raw	79.8	7-75	2.3	16.7	.8	1.1	.1	
Kale:								
Leaves, without stems (midribs)	82.7	53	6.0	9.0	—	1.5	.8	Calories increase with storage. High in calcium.**

Amounts in 100 grams, edible portion

VEGETABLES

FOOD AND DESCRIPTION	WATER	FOOD ENERGY	PRO-TEIN	CARBOHYDRATE TOTAL	CARBOHYDRATE FIBER	ASH	TOTAL FAT	MISCELLANEOUS AND COMMENTS
	percent	calories	grams	grams	grams	grams	grams	
Leaves, including stems	87.5	38	4.2	6.0	1.3	1.5	.8	
Cooked:								
Leaves, without stems (midribs)	87.8	39	4.5	6.1	—	.9	.7	
Leaves, including stems	91.2	28	3.2	4.0	1.1	.9	.7	
Kohlrabi, bulb:								
Raw	90.3	29	2.0	6.6	1.0	1.0	.1	
Cooked	92.2	24	1.7	5.3	1.0	.7	.1	
Leeks, bulb and lower leaf portion, raw	85.4	52	2.2	11.3	1.3	.9	.3	Mild substitute for onions.
Lentils, mature seeds, dry:								*
Whole:								
Raw	11.1	340	24.7	60.1	3.9	3.0	1.1	
Cooked	72.0	106	7.8	19.3	1.2	.9	trace	
Split, without seed coat, raw	10.4	345	24.7	61.8	2.7	2.2	.9	
Lettuce, raw:								
Butterhead varieties, as Boston types and Bibb	95.1	14	1.2	2.5	.5	1.0	.2	
Cos or romaine, as Dark Green and White Paris	94.0	18	1.3	3.5	.7	.9	.3	
Crisphead varieties, as Iceberg, New York and Great Lakes strains	95.5	13	.9	2.9	.5	.6	.1	
Loose leaf or bunching varieties, as Grand Rapids, Salad Bowl, Simpson	94.0	18	1.3	3.5	.7	.9	.3	
Mushrooms:								
Agaricus campestris, cultivated commercially:								
Raw	90.4	28	2.7	4.4	.8	.9	.3	

continued

Amounts in 100 grams, edible portion

VEGETABLES

FOOD AND DESCRIPTION	WATER	FOOD ENERGY	PRO-TEIN	CARBOHYDRATE TOTAL	CARBOHYDRATE FIBER	ASH	TOTAL FAT	MISCELLANEOUS AND COMMENTS
	percent	calories	grams	grams	grams	grams	grams	
Mushrooms, continued								
Canned, solids and liquid	93.1	17	1.9	2.4	.6	1.6	.1	
Other edible species, raw	89.1	35	1.9	6.5	1.1	1.0	.6	
Mustard greens:								
Raw	89.5	31	3.0	5.6	1.1	1.4	.5	
Cooked	92.6	23	2.2	4.0	.9	.8	.4	
New Zealand spinach:								Contains large amounts of oxalic acid.**
Raw	92.6	19	2.2	3.1	.7	1.8	.3	
Cooked	94.8	13	1.7	2.1	.6	1.2	.2	
Okra:								
Raw	88.9	36	2.4	7.6	1.0	.8	.3	
Cooked	91.1	29	2.0	6.0	1.0	.6	.3	
Onions, mature (dry):								Raw—Used as antiseptic.
Raw	89.1	38	1.5	8.7	.6	.6	.1	
Cooked	91.8	29	1.2	6.5	.6	.4	.1	
Onions, young green (bunching varieties), raw:								
Bulb and entire top	89.4	36	1.5	8.2	1.2	.7	.2	
Bulb and white portion of top	87.6	45	1.1	10.5	1.0	.6	.2	
Tops only (green portion)	91.8	27	1.6	5.5	1.3	.7	.4	
Onions, Welsh, raw	90.5	34	1.9	6.5	1.0	.7	.4	
Parsley, common garden and curled-leaf varieties, raw	85.1	44	3.6	8.5	1.5	2.2	.6	High in Vitamin C.**
Parsnips:								
Raw	79.1	76	1.7	17.5	2.0	1.2	.5	
Cooked	82.2	66	1.5	14.9	2.0	.9	.5	

Amounts in 100 grams, edible portion

VEGETABLES

FOOD AND DESCRIPTION	WATER	FOOD ENERGY	PRO-TEIN	CARBOHYDRATE TOTAL	FIBER	ASH	TOTAL FAT	MISCELLANEOUS AND COMMENTS
	percent	calories	grams	grams	grams	grams	grams	
Peas, edible-podded:								
Raw	83.3	53	3.4	12.0	1.2	1.1	.2	
Cooked	86.6	43	2.9	9.5	1.2	.8	.2	
Peas, green, immature:								
Raw	78.0	84	6.3	14.4	2.0	.9	.4	
Cooked	81.5	71	5.4	12.1	2.0	.6	.4	
Canned, solids and liquid:								
Alaska (Early or June peas)	82.6	71	3.5	12.5	1.5	1.1	.3	
Sweet (sweet wrinkled peas, sugar peas)	84.8	57	3.4	10.4	1.4	1.1	.3	
Peas, mature seeds, dry:								
Whole, raw	11.7	340	24.1	60.3	4.9	2.6	1.3	*
Split, without seed coat:								
Raw	9.3	348	24.2	62.7	1.2	2.8	1.0	
Cooked	70.0	115	8.0	20.8	.4	.9	.3	
Peppers, hot chili:								Dried hot pods are extremely high in Vitamin A.
Immature green, raw pods, excluding seeds	88.8	37	1.3	9.1	1.8	.6	.2	
Canned pods, excluding seeds, solids and liquid	92.5	25	.9	6.1	1.2	.4	.1	
Mature Red:								
Raw:								
Pods, including seeds	74.3	93	3.7	18.1	9.0	1.6	2.3	
Pods, excluding seeds	80.3	65	2.3	15.8	2.3	1.2	.4	
Dried pods	12.6	321	12.9	59.8	26.2	7.4	9.1	
Peppers, sweet, garden varieties:								Raw, sweet red are among the highest in Vitamin C.
Immature green:								
Raw	93.4	22	1.2	4.8	1.4	.4	.2	
Cooked	94.7	18	1.0	3.8	1.4	.3	.2	
Mature, red, raw	90.7	31	1.4	7.1	1.7	.5	.3	

continued

219

Amounts in 100 grams, edible portion

VEGETABLES

FOOD AND DESCRIPTION	WATER	FOOD ENERGY	PRO-TEIN	CARBOHYDRATE TOTAL	CARBOHYDRATE FIBER	ASH	TOTAL FAT	MISCELLANEOUS AND COMMENTS
	percent	calories	grams	grams	grams	grams	grams	
Pickles, cucumber:								
Dill	93.3	11	.7	2.2	.5	3.6	.2	
Sour	94.8	10	.5	2.0	.5	2.5	.2	
Sweet	60.7	146	.7	36.5	—	1.7	.4	
Pigeon peas, raw:								
Immature seeds	69.5	117	7.2	21.3	3.3	1.4	.6	
Mature seeds	10.8	342	20.4	63.7	7.0	3.7	1.4	*
Pimientos, canned, solids and liquid	92.4	27	.9	5.8	.6	.4	.5	
Potatoes:								
Raw	79.8	76	2.1	17.1	.5	.9	.1	High in carbohydrates.
Cooked:								
Baked in skin	75.1	93	2.6	21.1	.6	1.1	.1	
Boiled in skin	79.8	76	2.1	17.1	.5	.9	.1	
Canned, solids and liquid	88.5	44	1.1	9.8	.2	.4	.2	
Pumpkin:								
Raw	91.6	26	1.0	6.5	1.1	.8	.1	
Canned	90.2	33	1.0	7.9	1.3	.6	.3	
Pumpkin and squash seed kernels, dry	4.4	553	29.0	15.0	1.9	4.9	46.7	High in protein and calories.
Radish, raw:								
Common	94.5	17	1.0	3.6	.7	.8	.1	
Oriental, including daikon	94.1	19	.9	4.2	.7	.7	.1	
Rhubarb:								
Raw	94.8	16	.6	3.7	.7	.8	.1	Contains large amounts of oxalic acid.
Cooked	62.8	141	.5	36.0	.6	.6	.1	
Rutabagas:								
Raw	87.0	46	1.1	11.0	1.1	.8	.1	
Cooked	90.2	35	.9	8.2	1.1	.6	.1	

Amounts in 100 grams, edible portion

VEGETABLES

FOOD AND DESCRIPTION	WATER	FOOD ENERGY	PRO-TEIN	CARBOHYDRATE TOTAL	CARBOHYDRATE FIBER	ASH	TOTAL FAT	MISCELLANEOUS AND COMMENTS
	percent	calories	grams	grams	grams	grams	grams	
Salsify:								
Raw	77.6	13-82	2.9	18.0	1.8	.9	.6	
Cooked	81.0	13-82	2.6	15.1	1.8	.7	.6	
Sauerkraut, canned, solids and liquid	92.8	18	1.0	4.0	.7	2.0	.2	
Seaweeds, raw:								Very high in iodine. Also high in sodium and potassium.
Agar	16.3	—	—	—	.7	3.7	.3	
Dulse	16.6	—	—	—	1.2	22.4	3.2	
Irish moss	19.2	—	—	—	2.1	17.6	1.8	
Kelp	21.7	—	—	—	6.8	22.8	1.1	
Laver	17.0	—	—	—	3.5	11.0	.6	
Shallots, bulbs, raw	79.8	72	2.5	16.8	.7	.8	.1	
Soybeans:								Dried soybeans are highest of all vegetables in protein.
Immature seeds:								
Raw	69.2	134	10.9	13.2	1.4	1.6	5.1	*
Cooked	73.8	118	9.8	10.1	1.4	1.2	5.1	
Canned, solids and liquid	81.8	75	6.5	6.3	.7	2.2	3.2	
Mature seeds, dry:								
Raw	10.0	403	34.1	33.5	4.9	4.7	17.7	
Cooked	71.0	130	11.0	10.8	1.6	1.5	15.7	
Fermented products:								Fermented products and sprouted seeds contain digestive enzymes.
Natto (soybeans)	62.7	167	16.9	11.5	3.2	1.5	7.4	
Miso (soybeans and cereal)	53.0	171	10.5	23.5	2.3	8.4	4.6	
Sprouted seeds:								
Raw	86.3	46	6.2	5.3	.8	.8	1.4	
Cooked	89.0	38	5.3	3.7	.8	.6	1.4	
Soybean curd (tofu)	84.8	72	7.8	2.4	.1	.8	4.2	
Soy sauce, Tamari	62.8	68	5.6	9.5	0	20.8	1.3	High in sodium.

continued

221

Amounts in 100 grams, edible portion
VEGETABLES

FOOD AND DESCRIPTION	WATER	FOOD ENERGY	PROTEIN	CARBOHYDRATE TOTAL	CARBOHYDRATE FIBER	ASH	TOTAL FAT	MISCELLANEOUS AND COMMENTS
	percent	calories	grams	grams	grams	grams	grams	
Spinach:								Contains large amounts of oxalic acid. High in zinc. **
Raw	90.7	26	3.2	4.3	.6	1.5	.3	
Cooked	92.0	23	3.0	3.6	.6	1.1	.3	
Canned, solids and liquid	93.0	19	2.0	3.0	.7	1.6	.4	
Squash:								Winter varieties contain large amounts of Vitamin A.
Summer, all varieties:								
Raw	94.0	19	1.1	4.2	0.6	0.6	0.1	
Cooked	95.5	14	.9	3.1	.6	.4	.1	
Zucchini and Cocozelle, green:								
Raw	94.6	17	1.2	3.6	.6	.5	.1	
Cooked	96.0	12	1.0	2.5	.6	.4	.1	
Winter, all varieties:								
Raw	85.1	50	1.8	12.4	1.4	.8	.3	
Cooked:								
Baked	81.4	63	1.1	15.4	1.8	1.0	.4	
Boiled, mashed	88.8	38	1.1	9.2	1.4	.6	.3	
Butternut:								
Raw	83.7	54	1.4	14.0	1.4	.8	.1	
Cooked:								
Baked	79.6	68	1.8	17.5	1.8	1.0	.1	
Boiled, mashed	87.8	51	1.1	10.4	1.4	.6	.1	
Sweet potatoes:								Higher in carbohydrates than white potatoes. Very high in Vitamin A.***
Raw:								
Firm-fleshed (Jersey types)	74.0	102	1.8	22.5	.9	1.0	.7	
Soft-fleshed (mainly Porto-Rico variety)	69.7	117	1.7	27.3	.7	1.0	.3	
Cooked:								
Baked in skin	63.7	141	2.1	32.5	.9	1.2	.5	
Boiled in skin	70.6	114	1.7	26.3	.7	1.0	.4	
Canned, liquids and solid, without added sugar or salt	88.0	46	.7	10.8	.3	.4	.1	
Tomatoes, green, raw	93.0	24	1.2	5.1	.5	.5	.2	

Amounts in 100 grams, edible portion

VEGETABLES

FOOD AND DESCRIPTION	WATER	FOOD ENERGY	PRO-TEIN	CARBOHYDRATE TOTAL	CARBOHYDRATE FIBER	ASH	TOTAL FAT	MISCELLANEOUS AND COMMENTS
	percent	calories	grams	grams	grams	grams	grams	
Tomatoes, ripe:								
Raw	93.5	22	1.1	4.7	.5	.5	.2	Rich in Vitamin C.
Cooked, boiled	92.4	26	1.3	5.5	.6	.6	.2	
Canned, solids and liquid	93.7	21	1.0	4.3	.4	.8	.2	
Turnips:								
Raw	91.5	30	1.0	6.6	.9	.7	.2	
Cooked	93.6	23	.8	4.9	.9	.5	.2	
Turnip greens, leaves, including stems:								
Raw	90.3	28	3.0	5.0	.8	1.4	.3	Contains large amounts of oxalic acid.**
Cooked in small amount of water, short time	93.2	20	2.2	3.6	.7	.8	.2	
Watercress, leaves, including stems, raw	93.3	19	2.2	3.0	.7	1.2	.3	
Yam, tuber, raw	73.5	101	2.1	23.2	.9	1.0	.2	Higher in carbohydrates than white potatoes.***

* Dried beans highest in protein, potassium and calories of all vegetables.
** All leafy green vegetables are high in Vitamin A.
*** Sweet potatoes contain larger amounts of Vitamin A than yams, but yams contain more protein.

SMALL-GARDEN VEGETABLE GROWING CHART

CROP	PLACE TO GROW	TIME OF YEAR	WHERE TO START	WHEN TO START	HOW LONG	AVERAGE SIZE AT MATURITY
Asparagus	Buy as 1-year-old plant or seeds and plant in garden Plant 1" deep	Early spring	Soak seed in warm water for 2 or 3 days before planting	Early spring	1st year—no cutting 2nd year—light cutting, not extending more than 2 weeks In succeeding years—6-8 weeks to early July	3/8" in diameter 5½" long (No yield first year)
Beans, bush: snap or lima	8" to 10" of soil—Any container Plant 1½" to 2" deep	Warm weather	Directly into a container	Early spring	Lima: 65 days Snap: 50-55 days	1' to 2' in height
Beets	10" to 12" of soil—Any container Plant ½" deep	Early spring, fall Cool weather	Directly into a container	2-4 weeks before frost-free date	50-60 days	10" to 12" in height
Broccoli: early late	10" to 12" of soil. Transplant. Plant ½" deep. Any container	Early spring Mid-summer	1st year start seeds indoors, transplant when 4" high	Early spring Mid-summer	60-80 days	1st heads largest 3" to 6" in diameter 8" to 10" in height
Carrots	10" to 12" of soil—Any container Plant ½" deep	Early spring, fall Cool weather	Directly into a container	2-4 weeks before frost-free date	65-75 days	10" to 12" in height
Cucumbers	Single plant—hanging basket 1 plant—gallon-size container 3-4 plants—5-gallon size container Plant ½" to 1" deep	Warm weather	For early start—plant in peat pots	3-4 weeks before frost-free date	55-70 days	Controlled by pinching back vines

CROP	PLACE TO GROW	TIME OF YEAR	WHERE TO START	WHEN TO START	HOW LONG	AVERAGE SIZE AT MATURITY
Eggplant	12" to 14" diameter pot for each plant Not less than 3-gallon size Plant ½" to 1" deep	Warm weather	For early start—plant in peat pots	8-9 weeks before transplant time	120-140 days	2' to 3' in height
Green peppers	Not less than gallon size per plant	Warm weather	For early start—plant in peat pots	7-8 weeks before frost-free date	110-120 days	2' to 3' in height
Lettuce	Size determined by quantity Any container Plant ¼" deep	Cool weather Can withstand light frost	Directly into a container	4-6 weeks before frost-free date	40-50 days	6" to 10" in height
Onions, green	8" to 10" of soil—Any container Transplant plant seeds ½" deep	Cool weather Can withstand light frost	Directly into a container	2-6 weeks before frost-free date	35-45 days	10" to 12" in height
Radishes	6" of soil—Any container Plant ¼" deep	Cool weather Can withstand light frost	Directly into a container	2-4 weeks before frost-free date	24-30 days	6" to 8" in height
Spinach	8" to 10" container per plant Plant ½" deep	Early spring, fall Cool weather	Directly into a container	2-4 weeks before frost-free date	50-70 days	Only a few inches tall Growth is in width
Squash	Grow on mounds in 5-gallon container for 3-4 plants; i.e., bushel basket, washtub, etc. Plant 1" deep	Warm weather—until fall	For early start—plant in peat pots	3-4 weeks before frost-free date	50-60 days summer 85-110 days winter	Bush—2' to 3' in height Vine—Pinch off ends of vines to control
Tomatoes	Miniatures—8" to 10" pot Standard—2-3 gallon size Dwarf—gallon size Transplant	Warm weather	For early start—plant seedlings in peat pots	6-8 weeks before frost-free date	50-90 days; depends on type grown	Standard—3' to 5' in height Dwarf—2' to 3' in height

Acid and Alkaline Content

All digested food leaves either an acid or an alkaline ash in the body. The designation of foods as acid or alkaline is the modern scientific counterpart of the oriental system of opposites, yin and yang. The body's living tissues are slightly alkaline, and the waste matter slightly acid. People who are extremely alkaline tend to be tense, rigid, and slow; those who are highly acid usually have excessive nervous energy. It is essential to maintain a balance; either extreme is dangerous. The more mineral-rich food one eats, the better the chance there is for proper stability. As the following charts demonstrate (For fruit charts see p. 260; for nut charts see p. 266.) most vegetables and fruit provide necessary alkalinity (meat generally produces an acid ash).

VEGETABLES

Yin and Yang

Yang △
More Yang △△

Yin ▽
More Yin ▽▽
Very Yin ▽▽▽

Artichoke ▽▽▽
Asparagus ▽▽▽
Beans ▽▽▽ (except azuki)
Beets ▽
Burdock △△
Cabbage (white) ▽
Cabbage (red) ▽▽
Carrot △△
Celery ▽▽
Corn ▽

Cucumber ▽▽▽
Dandelion leaves △
Dandelion roots △△▽▽▽
Eggplant ▽▽▽
Endive ▽▽
Garlic ▽▽
Kale △
Lentil ▽▽
Lettuce △
Mushroom ▽▽▽
Olive ▽

Onion △
Parsley △
Peas, (green) ▽▽
Potato ▽▽▽
Pumpkin △△
Radish ▽▽▽
Spinach ▽▽
Sweet potato ▽▽▽
Tomato ▽▽▽
Turnip △

Acid and Alkaline Content of Vegetables

Alkaline Ash

Asparagus
Beans, lima
Beans, navy, pea
Beans, snap
Beets
Cabbage, cooked
Carrots
Cauliflower
Celery
Chard, Swiss

Cucumber
Dandelion greens
Eggplant
Endive, curly
Kale
Kohlrabi
Lettuce
Mushrooms
Okra
Olives, green and ripe

Onions
Parsnips
Peas
Peppers
Potato, white
Potato, baked
Pumpkin
Radish
Rutabagas
Salsify

Sauerkraut
Squash, summer
Squash, winter
Sweet potato
Tomatoes
Turnip
Turnip greens
Watercress

Acid Ash

Corn
Lentils, dried
Rhubarb

227

Fruit Glossary

APPLES

Consumer information Available all year. There are a large variety of apples, some of which are more suited for cooking and others usually eaten raw. Their appearance, taste, and availability varies. Basically the Gravenstein (July-September) are tart and good for pies; the Grimes Golden (September-December) are good for cooking; the Jonathan (September-May) are good for dessert, cooking, baking; the Wealthy (August-October) good for baking, cooking, sauce, etc.; the Red Delicious (September-May) are good in salads and for eating out of hand, and they are generally not cooked; the Golden Delicious (October-June) are good in salads, for eating out of hand, etc; the McIntosh (October-May) is an all-purpose apple; the Rome Beauty (October-April) is good for pies, sauce, and baking, but is generally not eaten raw; and the Winesap (November-July) is an all-purpose apple. When buying, look for fresh, crisp, firm, unbruised apples. They should be well colored for their variety.
Note: Dried apples are available in health stores and some supermarkets and make a nutritious snack. Applesauce, canned cooked apples, etc., are also available all year.

Storing If apples are still hard when bought, let them ripen at room temperature. Otherwise, they should be put in the refrigerator. In a cool place, apples may be stored for several weeks or even months.

Uses Eaten raw (for varieties, see above), use in fruit salads, baked, cut and baked in cakes, pies, etc., cored and stuffed and baked whole, fried, broiled, pureed, spiced, preserved (apple butter, jelly, etc.), as garnish, included in chutneys, made into vinegars, etc.

Tip Dip in fresh lemon juice to avoid discoloring.

APRICOTS

Consumer information Available fresh mostly in June and July. Available all year canned, dried, frozen. Look for plump, well-formed, fairly firm fruit with deep uniform yellow or yellowish orange color. Avoid the yellowish green, unripened ones because they do not gain sugar after they are picked and will not ripen in your kitchen. Also avoid overripened fruits, which tend to be soft and mushy.

Storing   In refrigerator for two or three days.

Uses   Wash and eat fresh apricots. It is not necessary to peel them. Use in fruit salads, as meat garnish, in desserts, as appetizers, in baked goods, mousses, soufflés, and sherbets. Eat dried apricots directly as snacks, or chop and put in cakes, cookies, etc.; cook in water and serve stewed; soak and use as meat garnish. Apricots also can be pureed, spiced, brandied, candied, and preserved.

AVOCADOS

Consumer information   Available all year. Look for fruit that is heavy for its size, bright and fresh-appearing, not shriveled or bruised. The color ranges from purple-black to green according to variety. The size also varies, as does the skin, which may be thin and smooth, thick and smooth, or leathery and rough. While it is all right to buy avocados when they are still firm and hard, they should be eaten when ripe. A test for ripeness is when the flesh yields to gentle pressure of the fingers.

Storing   Eat as soon as possible after ripening at room temperature. Avoid refrigeration, but if absolutely necessary store in refrigerator for 3 to 5 days.

Uses   Cut in halves or smaller pieces, then peel. Slice, chop, or mash. Serve in salads with vegetables or fruits, alone *à la vinaigrette,* mashed in guacamole or as a dip, stuffed and baked, pureed as an appetizer, hot or cold in soups and sauces, jellied in molds.

Tip   Dip in fresh lemon juice to avoid discoloring.

BANANAS

Consumer information   Available all year round, although supply is higher in May and June than in other months. There are two main varieties, the Gros Michel and the Cavendish. The Gros Michel is the most frequent variety seen on the markets and is more tapered at the ends, while the Cavendish is blunter. When buying, look for plump, well-formed fruit. They should not be bruised or split. The color may vary from green to yellow with brown spots, depending on degree of maturity. Bananas continue to ripen off the tree. In fact, bananas that are picked before maturity and that ripen off the tree tend to have more flavor. It is important to know that the truly ripe banana has a golden yellow skin, flecked with brown spots, and should be eaten raw only in this state. When green or partially ripe, they must be cooked, because the banana starch has not been fully converted into digestible fruit sugars.

Note: Also available dried in natural food stores and in flour form.

Storing To ripen, leave at room temperature until yellow flecked with brown spots. Refrigeration is inadvisable, since the low temperatures impair the flavor and prevent proper normal ripening. If a banana becomes overripe, it may be used in a number of ways, listed below.

Uses Ripe fruit is peeled and eaten raw, sliced and served with cereals, added to fruit salads, served along with main course as side dish. Bananas that are not thoroughly ripe should be cooked, baked, fried, etc., and served as garnish, or as a dessert (fried and flambéed). An overripe banana may be mashed and included in cakes, pie fillings, breads, muffins, pudding, milk shakes, etc. Bananas may also be dried, pounded into flour, made into vinegar, and preserved.

BERRIES (BLACKBERRIES, BOYSENBERRIES, LOGANBERRIES, RASPBERRIES, ETC.)

Consumer information Available May through August, with peak in June and July. Some berries are available frozen all year round. Look for bright characteristic color, fresh, clean, dry appearance. They should be free of mold and bruises and should not show signs of leaking. Check for staining on the cartons, which will indicate crushed and leaky fruit. Overripe berries will have a dull color and be oversoft. Unripe berries will show green or off-color drupelets.

Storing All berries are fragile and perishable. Therefore, they should be eaten right away and refrigerated only for a short time. Always make sure that nothing is placed on top of them or presses against them.

Uses   Eat fresh as topping for cereal, in fruit salads, or with cream and sugar; cook in tarts, pies, cakes, pancakes. If they become soft, mash them and cook briefly in simple sugar syrup to make a topping for pancakes, ice cream, etc.; also make into ice cream; juice; preserve in jams, etc.

BLACKBERRIES (see Berries)

BLUEBERRIES

Consumer information   Available May through August, with the peak in June and July. Look for fresh, plump berries that are uniform in size. They may have a waxy bloom, depending on variety. They should have a good blue appearance. Old berries will be dull in color and shriveled. Refuse moldy or leaky berries.

Storing   Blueberries are fragile and perishable and should be refrigerated and eaten right away. Blueberries can be easily frozen and enjoyed all year. Buy a quantity of fresh fruit and put in plastic containers or bags, seal, and place in freezer. Some commercial frozen blueberries can also be purchased.

Uses   Fresh berries are served on cereals or ice creams, in mousses and blended drinks, with cream and sugar; baked in pies, cakes, tarts, muffins, and pancakes. They may be cooked into syrups and toppings, preserved, or pickled. (Pickle berries with molasses, cider vinegar, honey, salt, and nutmeg and serve as garnish.)

BOYSENBERRIES (see Berries)

CHERRIES

Consumer information   Available fresh May through July. Buy only cherries that are fresh, firm, and highly colored (ranging from red to black, depending on variety). Fruit should not be sticky, leaky, or bruised. Immature cherries are hard, too light in color for their variety, generally small. Unripe cherries will taste acid and dry and may also cause stomach disorders. The best method for choosing cherries is the taste method. You will know the ripe fruit by its juicy, fine flavor.

Note: Dried cherries are available in natural-food stores and in oriental grocery stores and are a nourishing snack.

Storing   Cherries do not store well. They should be bought and eaten as quickly as possible. They may be refrigerated for a short period.

Uses   Eat fresh, pitted and spiced, brandied, pickled, preserved; pitted and baked in pies, made into sauces (e.g., for duck), stewed, juiced, pureed. Sour cherries can be used to make a soup (served cold in summer and warm in winter). Chocolate-covered cherries are sold in candy stores. Maraschino cherries are used in many beverages. Kirchwasser is a very popular liqueur made from cherries and cherry pits. Cherries are also used in a variety of drugs, e.g., coughdrops.

COCONUTS

Consumer information   Available all year but mostly September through December. All coconuts sold on the market are ripe. Select a nut that is heavy for its size and full of liquid; it should make a sloshy sound when shaken. If the nut is dry, it has spoiled. Nuts with moldy or wet eyes are also not good.

Storing   Refrigerate.

Uses   Shell in either of two ways: drain milk and place in freezer for approximately an hour or put in moderate oven (350° F.) briefly. Then rap sharply with hammer and shell will shatter. Eat fresh coconut meat, or use dried and pounded into meal; dessicated shredded coconut is used to make a pudding (when added to milk and arrowroot). Coconut milk is heated and used as a sauce or addition to chutneys, mashed yams, potatoes, curries, etc. The milk is also added to beverages.

CRANBERRIES

Consumer information   Available fresh September through December, but all year frozen, jellied, juiced, and in relishes. Most cranberries on the market are of good quality, so selecting them is usually easy. Look for fresh, plump, firm berries. The color should be red to reddish black. Poor-quality berries will be shriveled, dull in color, soft, and sticky.

Storing Refrigerate and use within a week to ten days. Cranberries are easily frozen without any processing. Simply put them in plastic container and place in freezer.

Uses Baked in pies, breads, and muffins. Used in jellied salads, added to mousses, chopped into relishes and sauces, used as a garnish for meats and poultry, juiced, made into sherbet, etc.

DATES

Consumer information Dried dates are available all year. Look for golden brown fruit that is plump and soft. Dates are available in grocery stores and in specialty shops.

Storing After package has been opened, refrigerate. Keep dates well wrapped to avoid drying and hardening. Do not store them near onions, fish, or other strong-smelling foods, because they tend to absorb odors.

Uses Bake in breads, cakes, pies, or waffles. Add to puddings. Make into spreads for sandwiches (e.g., with cream cheese or with peanut butter). Chop and blend with milk for a beverage. Eat straight from package as snack. Pit and stuff with nuts, etc.

FIGS

Consumer information Available fresh in August, September, and October and all year round dried. When buying fresh figs, look for full, soft-ripe fruit with characteristic green-yellow to purple-black color, depending on variety. An overripe fig will have a sour odor, due to fermentation of juice. Watch for and avoid damaged or moldy fruit. When buying dried figs, look for either black or golden brown color. Figs are also available in cans.

Storing Fresh figs are extremely perishable and should be bought for immediate use. Since they bruise easily, place carefully on fruit platter or in refrigerator and avoid crowding the fruit.

Uses Eat fresh (peel and eat meat inside). Use dried figs in breads, muffins, candies, confections, cookies, puddings, mousses, desserts, etc. Use in main entrées, sandwiches, sauces, or as a garnish.

GRAPEFRUIT

Consumer information Available October through June. Pick grapefruit that are firm, springy to the touch, and heavy for their size. Avoid ones that are soft, loose-skinned, and irregularly shaped. Heavy fruits tend to be thin-skinned and contain more juice than thick-skinned varieties. Fruits somewhat pointed at the stem end tend to be thick-skinned. Russeting does not harm the fruit, nor does the bronze color indicate an overripe fruit. The white grapefruit meat tends to have a stronger flavor than the pink. Select size according to need, i.e., large fruits when segmenting the fruit and medium-sized for serving in halves.

Storing Fruit may be kept at room temperature or refrigerated.

Uses Eat fresh by peeling or halving and eating fruit with spoon; or segmenting and adding to fruit salads, appetizers, etc. Juice, use in sauces, chutneys, compotes, etc. Broil with brown sugar and cinnamon. Candy the peels.

GRAPES

Consumer information Available most of the year. Grapes do not ripen off the vine. A ripe grape should be firmly attached to its stem and the fruit should appear fresh, smooth, plump, and not sticky. They should have a vivid characteristic color. Avoid overdry stems or leaky fruit.

Storing Refrigerate.

Uses In fruit salads, eaten off stem, jellied, preserved, as garnish for main course, cooked with chicken, meat, vegetables. In French cuisine the term *véronique* refers to dishes cooked with grapes. Grapes can be juiced, and fermented juice is used for wine.

Tip   A regimen of grapes or grape juice is used as a cleansing diet.

GUAVAS

Consumer information   Rarely found on the market fresh. However, they are round, oval, oblong, or pear-shaped, with a thin skin colored green to bright yellow, sometimes with a pink blush.

Storing   Refrigerate.

Uses   Preserves, jellies, etc. Often cooked and canned. Made into a paste or "cheese" by boiling. Guava vinegar is sometimes used for people who have difficulty digesting other vinegars.

LEMONS

Consumer information   Available all year. Look for lemons that have fine-textured skin and are heavy for their size and moderately firm. They should have a rich yellow color with no traces of green. Since they are picked in a green state and ripened off the tree, it is best to look for ripe fruit that is yellow and springy to the touch.

Storing   Lemons are better kept at room temperature than refrigerated.

Uses   Juiced and used as a beverage; added to other beverages such as tea; served with fish, veal, chicken, etc. Used in sauces, as garnish, in chutneys and relishes, in mousses, pie fillings, puddings, etc. Used to tenderize veal, squeezed on melons, fruit salads, vegetable salads. The rind is used in baking and can also be candied.

LIMES

Consumer information   Available all year. Peak season in June through August. Domestic limes are bright green and heavy for their size. Yellow fruits lack acidity. Fruits with purple-brown spots are undesirable. The Mexican variety, Key lime, is bright yellow when ripe and is very flavorful.

Storing At room temperature or refrigerated.

Uses Juiced and added to beverages, sauces, pie fillings, meats, fish, poultry, etc.

LOGANBERRIES (see Berries)

MANGOES

Consumer information Available May through August, with peak in June. Look for smooth outer skin, usually green with yellowish or red areas. The red and yellow increases as the fruit ripens. The pulp is yellow and delicate in flavor (something between an apricot and a pineapple in flavor). When unripe, it tastes acid and unpleasant. The shape is round to oval and the size varies. Avoid wilted or discolored and decaying fruit.

Storing Store at room temperature until soft. Then refrigerate and eat as soon as possible.

Uses In fruit cups and salads, or peeled and eaten alone. Baked in cakes, canned, frozen, preserved. In jams, chutneys (unripe fruits are best for this), purees, curries, sauces, and ice creams. Powdered dried mango (amchoor) is used as flavoring in vegetable dishes. Dried seeds are used in the Far East, flowers are eaten in Siam, leaves in Java.

MELONS

Cantaloupe

Consumer information In major cities, available all year. However, main season is May through September, with peak period in June through August. Look for smooth, rounded melon, with depressed scar at stem end. This indicates that melon was picked at "full slip"—that is, when ripe.

A clue to an unripe melon is pieces of stem that may still be attached. Not fully ripe melons can still be sweet because they develop all their sugar content on the vine.

Storing If the melon is not soft when you buy it, allow it to soften at room temperature for a few days, during which time it will take on a distinct aroma and a yellowish appearance. When ripe, refrigerate.

Uses Eat fresh, halved or quartered, with seeds scooped out. Cut into fruit salads, or fill with other fruits, cottage cheese, crab meat, etc. As an appetizer, wedges may be wrapped with prosciutto (smoked ham); serve as dessert with sherbet.

Casaba

Consumer information Available July through November, with peak in September and October. A ripe melon is indicated by yellow rind and slight softening at stem end. The flesh of a ripe melon will be soft, creamy white, sweet, and juicy.

Storing At room temperature if not yet fully ripe. Refrigerate if fully ripe. A casaba will ripen off the vine.

Uses Same as for cantaloupes.

Crenshaw

Consumer information Available July through October, with peak in August and September. A ripe melon will show softening of the rind, especially at large end, and will have a golden skin and rich aroma.

Storing At room temperature if not fully ripe. Refrigerate if fully ripe.

Uses Generally serve in wedges because melon is large in scale and a wedge will equal a serving. Other uses, same as for cantaloupes.

Honeydew

Consumer information Available July through October. Select ripe melon by avoiding any melons that are dead white with greenish tint. Generally, a creamy white skin color indicates ripeness. Also, if rind has hard, smooth feel rather than soft velvety feel, the melon is unripe and will not ripen off the vine.

Storing At room temperature preferably, because it improves flavor.

Uses Serve in wedges as this melon is very large. Uses are the same as for cantaloupes.

Persian

Consumer information Available July through October, with peak in August and September. Ripe fruit will be indicated by a slightly lighter green color of the rind under the netting of the surface. This melon will not ripen after being picked. The rind should give slightly under light pressure.

Storing At room temperature.

Uses Same as for cantaloupes.

Spanish (see *Crenshaw*).

Watermelons

Consumer information Available May into September. Look for ripe melon, which will be firm, symmetrical, fresh-looking, with waxy bloom. The lower side (the one that has had contact with the soil) should be slightly yellowish rather than white or pale green.

Storing At room temperature or refrigerate. It will not sweeten after being picked from vine.

Uses Serve in wedges, cut into fruit salads, etc. The rind is often pickled, or cut and used in chutneys.

NECTARINES

Consumer information Available mostly in June, July, and August. Consumer considerations are the same as for peaches. Nectarines do not ripen off the tree, so avoid green fruit. Also avoid shriveled fruit, which indicates premature picking. Nectarine flesh may be red, white, or yellow.

Storing Same as for peaches.

Uses Same as for peaches.

OLIVES

Consumer information Green and black olives are available year round in jars and cans, pitted and unpitted and stuffed, in many sizes. "Greek" black olives, which are pickled and slightly shriveled, are sold in bulk.

Storing Open jars or cans should be refrigerated.

Uses As an appetizer, in salads, cooked in stews, paellas, chicken and meat dishes, chopped and mixed with cream cheese or cheddar cheese as sandwich spread. Green olives may be purchased already stuffed with pimiento or nuts or you may stuff them.

ORANGES

Consumer information Available all year, with largest supplies December through April. Select fruit that is firm and heavy and has a fine-textured skin. Color varies with variety, but it does not reflect the maturity of the orange. Therefore, a greenish orange may be as ripe as a golden one. Sometimes a golden orange may turn green later. The bright Florida oranges may have color added to satisfy the public and are stamped "color added." Russeted oranges are just as good as the brightly colored fruit of the same variety. Avoid spongy, lightweight oranges.

Storing Fruit may be kept at room temperature or refrigerated.

Uses Eat fresh by peeling or quartering. Segment and add to fruit salads; juice; use in sauces, chutneys, relishes, mousses, jellied molds, etc.; make into sherbet, marmalades, etc. Candy the peel.

PAPAYAS

Consumer information Available mainly in May and June. Select well-colored (at least half yellow and with little green), smooth-skinned, unbruised, and unshriveled fruit. Avoid overly large fruit because they tend to be less flavorful than the medium to small size. Papayas should be basically pear-shaped. They should yield slightly to pressure between the palms.

Storing Ripen at room temperature, then refrigerate and use promptly.

Uses Eat raw, juice, or add to fruit salads.

PEACHES

Consumer information Available May through October, with the peak in June, July, and August. It is most important to pick only mature peaches. Any trace of greenish color will indicate an unripe fruit. Green fruit will not ripen and will only shrivel and be valueless. Because peaches do not gain sugar after being picked, a green peach will be distasteful and hard to digest. Look for golden-colored fruit with well-rounded cheeks and a well-defined crease or fold. The reddish blush does not necessarily indicate a preferable fruit, since some varieties blush only a little. Available all year in cans; frozen; and dried in natural-food stores and gourmet shops.

Storing Store at room temperature and use immediately when soft enough to eat. If you see signs of brown rot, use immediately because peaches rot very quickly. Refrigerate fruit if soft and eat as soon as possible.

Uses Eat fresh, sliced and served with cream, chunked into fruit salads, broiled, or stuffed and baked. May be pickled, brandied, spiced, canned, frozen, dried, pureed, or distilled into liqueur or wine. Bake in pies, cakes, cobblers, crisps. Make into ice cream, mousses, etc. Juice or use in jellies and preserves. Also use as garnish for main courses and in sauces (with duck, lamb, etc.).

PEARS

Consumer information Bartlett pears (the large yellow variety with red blush), Anjous (the green to greenish yellow variety), and Bosc (the dark yellow overlaid with cinnamon-colored russet) are available mainly August through March. Look for firm, blemish-free, well-shaped pears. A green pear will ripen at room temperature. In fact, most pears on the market are ripened off the tree. Available all year in cans (cooked), and in dried form in health-food stores and gourmet shops.

Storing Store at room temperature until slightly soft; then refrigerate. Avoid oversoftening. Do not crowd in refrigerator because they bruise easily. Pears may be stored in a cool cellar for several weeks or even months.

Uses Bake (Anjou and Bosc varieties are better for this) and stuff with nuts, etc. Eat fresh whole or cut into fruit salads, cook in fruit compotes, etc. Add to other fruit juices and blend. Can and spice, etc.

PERSIMMONS

Consumer information Available mostly October through December. Select bright-orange or deep-red fruit with green cap still attached. Buy plump oblong fruit with smooth, highly colored, unbroken skin with no damage spots.

Storing Keep at room temperature until the fruit is as soft as jelly. Then eat immediately, as this is its most ripe and flavorful stage. (Do not eat before this stage, because the fruit contains an unpleasant acid called tannin.)

Uses Eat fresh, cooked, candied. Make into puddings, cakes, purees, etc.

PINEAPPLES

Consumer information Available all year, with peak in March through June. Select fruit that is heavy for its size and generally large. Look for freshness in the crown leaves, making sure that they are not dried out and brown. A small, compact leaf crown, in relation to the size of the fruit, indicates a well-developed fruit. The eyes of the pineapple should be flat and almost hollow. Some suggest the following "thump" test: tap the fruit with the forefinger and thumb. If the fruit makes a thumping sound, it is fresh. Pulling out the crown leaves is really not a reliable test, nor is color. Fragrance is a clue to ripeness. Avoid pineapples with discolored or soft spots. Pineapples will not ripen or sweeten after they are picked. Available all year in cans, frozen, juiced; or dried in health-food stores and gourmet shops.

Storing   Refrigerate and eat as soon as possible.

Uses   Juiced, alone or in punches; fresh in fruit cocktails, salads, desserts; baked in breads, cookies, cakes, pies; pickled; cooked in jams, sauces for meat, fish, or desserts. Use as tenderizer for meat, particularly ham.

PLANTAINS

Consumer information   Available all year in small quantities. Plantains may be purchased in unripe green state or in ripened state (golden), depending on intended use. Select as you would bananas (see *Bananas*).

Storing   At room temperature.

Uses   Always cook before eating. Green plantains are generally used for making plantain chips (fried in oil and seasoned with garlic salt). Generally the yellow ripe fruit is preferred for recipes such as mashed, baked, fried, broiled, or candied plantains. Also use golden plantains in custard, cakes, patties, and compotes.

PLUMS

Consumer information   Available June through September, with peak in July and August. Look for mature fruit, because immature fruit will not ripen after being picked and will be tart. Look for plump, clean, slightly soft, fully colored fruit. An overripe fruit will be too soft and will probably be leaky. Some varieties will be fully ripe when the color is yellowish green, others when the color is red, or purplish blue, or black. Look for softening at tip of fruit as an indication of maturity. Cooked plums are also sold in cans.

Storing   If not fully soft, keep at room temperature. However, if soft, store in refrigerator.

Uses Eat fresh, whole or sliced into fruit salads, compotes, jellied molds, etc. Bake into cakes, breads, puddings, pies. Plums make good jellies, jams, butters, preserves, sauces, etc. Freeze: cut fruit, measure 1 quart of cut fruit, add ¾ cup sugar and ¼ teaspoon ascorbic acid powder. Mix with plums, coating each piece. Then pack plums, including syrup that forms, in freezer containers. Cover, label, freeze. Also, cooked plums can be canned.

POMEGRANATES

Consumer information Available September through November, with peak month October. The pomegranate is about the size of an apple and has a thin, tough rind, which is pink or bright red. The flesh is crimson color, and the fruit inside is made up of numerous small seeds surrounded by a juicy spicy pulp. Avoid fruit that looks dry and hard.

Storing Refrigerate. The seeds may be placed in a large screw top jar and frozen.

Uses Eat out-of-hand; cut open or peel. You may chew the seeds or spit them out. The pulp makes a good juice. Seeds can be added to salads or used to garnish main course for taste and color.

PRUNES

Consumer information Available year round.

Storing If storing for a length of time, refrigerate.

Uses Eat plain, as a confection, stew in compote; soak and bake in cakes, puddings, etc. Prune juice is particularly effective as a natural laxative.

QUINCES

Consumer information The quince resembles a hard-fleshed yellow apple or pear. It is not generally found on the market, but does appear in specialty stores.

Storing   Treat as apples or pears.

Uses   In marmalade, jelly, and preserves. Cook with sugar and add as a flavoring to cooked apples, pears, etc. Make into candy, chutneys, etc.

RASPBERRIES (see Berries)

RHUBARB (see under Vegetables)

STRAWBERRIES

Consumer information   Available all year, with peak in May and June. Look for bright, red-colored, shiny berries that are free from moisture and mold. Stained containers will indicate that the berries are leaky and should be avoided. The green caps should be attached and bright in color. Poor-quality berries are usually misshapen, lacking in color, texture, and flavor. Brown caps also indicate inferior strawberries. Size does not correspond to quality and often large strawberries are flavorless. Most stores carry frozen strawberries as well as canned ones. When freezing strawberries yourself, you will find that some varieties freeze better than others and that it is necessary to experiment with a number of types.

Storing   Strawberries are highly perishable and should be refrigerated and eaten right away. Spread one layer deep for best storage. Wash just before eating and remove caps after washing so that water does not get into strawberries and dilute flavor and change the texture. Also, if you choose to sweeten them with sugar, sprinkle on only a half hour before serving so that the berries will not soften too much.

Uses   Eat fresh or dip in confectioners' sugar, sprinkle with sugar and lemon juice just before serving, serve with cream, yogurt, or sour cream. Bake in pies, shortcakes, muffins. Serve in fruit salads. Make into sauces, jams, or garnishes. Serve on top of cereal. Juice and add to other fruit juices. Add whole berries to wine, champagne, or punches. Make into mousses, ice cream, or sherbet, etc.

TANGERINES

Consumer information   Available November through March, but mostly November through January, with peak in December. Choose fruit that is deep orange or almost red in color. It should be heavy for its size. While puffiness is normal, since the skin is so easily removed, the fruit should not have soft, water-soaked areas.

Storing Keep in refrigerator until ready to use. The fruit is perishable and should be used as soon as possible.

Uses Peel and eat fresh; section and add to fruit salads. Make into dessert with shredded coconut. Use in drinks, etc.

TOMATOES (see under Vegetables)

WATERMELONS (see Melons)

COMPOSITION OF FRUITS

Amounts in 100 grams, edible portion

FRUITS FOOD AND DESCRIPTION	VIT. A	Vitamins								Minerals				
		THIA-MIN B_1	RIBO-FLAVIN B_2	NIACIN	PYRI-DOXINE B_6	FOLIC ACID	VIT. C	CAL-CIUM	PHOS-PHORUS	IRON	SODIUM	POTAS-SIUM	MAG-NESIUM
	Internat'l Units	Mg.	Mg.	Mg.	Micro.	Micro.	Mg.	Mg.	Mg.	Mg.	Mg.	Mg.	Mg.
Apples:													
Raw common varieties:													
Freshly harvested and stored, not pared	90	.03	.02	.1	30	2.0	4	7	10	0.3	1	110	8
Freshly harvested, not pared	90	.03	.02	.1	30	2.0	7	7	10	.3	1	110	—
Stored, not pared	90	.03	.02	.1	30	2.0	3	7	10	.3	1	110	—
Dried:													
Uncooked	—	.06	.12	.5	—	—	10	31	52	1.6	5	569	22
Cooked, without added sugar	—	.01	.03	.5	—	—	trace	9	15	.5	1	162	6
Apricots:													
Raw	2,700	.03	.04	.6	71	3.3	10	17	23	.5	1	281	12
Canned, solids and liquid, water pack, without added sweetener	1,830	.02	.02	.4	54	.5	4	12	16	.3	1	246	7
Dried:													
Uncooked	10,900	.01	.16	3.3	—	4.7	12	67	108	5.5	26	979	62
Cooked, without added sugar	3,000	trace	.05	1.0	—	—	3	22	35	1.8	8	318	20
Avocados, raw, all commercial varieties	290	.11	.20	1.6	610	30.0	15	10	42	.6	4	604	—
Bananas:													
Raw:													
Common	190	.05	.06	.7	320	9.7	10	8	26	.7	1	370	33
Red	400	.05	.04	.6	—	—	10	10	18	.8	1	370	—
Blackberries:													
Raw	140	.02	.02	.2	—	13.7	7	32	19	.9	1	170	30
Canned, solids and liquid, juice pack	150	.02	.03	.3	24	14.0	10	22	13	.6	1	115	—
Blueberries:													
Raw	100	.03	.06	.5	91	8.0	14	15	13	1.0	1	81	6
Canned, solids and liquid, water pack, without added sweetener	40	.01	.01	.2	39	4.2	7	10	9	.7	1	60	4

Amounts in 100 grams, edible portion

FRUITS	Vitamins							Minerals					
FOOD AND DESCRIPTION	VIT. A	THIA-MIN B_1	RIBO-FLAVIN B_2	NIACIN	PYRI-DOXINE B_6	FOLIC ACID	VIT. C	CAL-CIUM	PHOS-PHORUS	IRON	SODIUM	POTAS-SIUM	MAG-NESIUM
	Internat'l Units	Mg.	Mg.	Mg.	Micro.	Micro.	Mg.	Mg.	Mg.	Mg.	Mg.	Mg.	Mg.
Boysenberries: Canned, solids and liquid, water pack, without added sweetener	130	.01	.10	.7	—	—	7	19	19	1.2	1	85	—
Breadfruit, raw	40	.11	.03	.9	—	—	29	33	32	1.2	15	439	—
Cantaloupe, see Muskmelons													
Carambola, raw	1,200	.04	.02	.3	—	—	35	4	17	1.5	2	192	—
Cherimoya, raw	10	.10	.11	1.3	—	—	9	23	40	.5	—	—	—
Cherries: Raw:													
Sour red	1,000	.05	.06	.4	85	6.0	10	22	19	.4	2	191	14
Sweet	110	.05	.06	.4	—	6.5	10	22	19	.4	2	191	—
Canned, solids and liquid, water pack, without added sweetener:													
Sour red	680	.03	.02	.2	—	—	5	15	13	.3	2	130	—
Sweet	60	.02	.02	.2	157	3.0	3	15	13	.3	1	130	9
Coconut meat:													
Fresh	0	.05	.02	.5	44	27.6	3	13	95	1.7	23	256	46
Dried, unsweetened	0	.06	.04	.6	—	—	0	26	187	3.3	—	588	90
Coconut Milk (liquid extracted from mixture of grated coconut meat and water)	0	.03	trace	.8	—	—	2	16	100	1.6	—	—	—
Crabapples, raw	40	.03	.02	.1	—	—	8	6	13	.3	1	110	—
Cranberries, raw	40	.03	.02	.1	—	1.7	11	14	10	.5	2	82	8

continued

Amounts in 100 grams, edible portion

FRUITS

FOOD AND DESCRIPTION	Vitamins								Minerals					
	VIT. A	THIAMIN B_1	RIBOFLAVIN B_2	NIACIN	PYRIDOXINE B_6	FOLIC ACID	VIT. C	CALCIUM	PHOSPHORUS	IRON	SODIUM	POTASSIUM	MAGNESIUM
	Internat'l Units	Mg.	Mg.	Mg.	Micro.	Micro.	Mg.	Mg.	Mg.	Mg.	Mg.	Mg.	Mg.
Currants, raw:													
Black European	230	.05	.05	.3	—	—	200	60	40	1.1	3	372	—
Red and white	120	.04	.05	.1	—	—	41	32	23	1.0	2	257	15
Dates, domestic, natural and dried	50	.09	.10	2.2	153	24.7	0	59	63	3.0	1	648	58
Figs:													
Raw	80	.06	.05	.4	130	14.0	2	35	22	.6	2	194	20
Canned, solids and liquid, water pack, without added sweetener	30	.03	.03	.2	—	—	1	14	14	.4	2	155	—
Dried, uncooked	80	.10	.10	.7	175	32.0	0	126	77	3.0	34	640	71
Gooseberries:													
Raw	290	—	—	—	—	—	33	18	15	.5	1	155	9
Canned, solids and liquid, water pack, without added sweetener	200	—	—	—	—	—	11	12	10	.3	1	105	—
Granadilla, purple (passion fruit) pulp and seeds, raw	700	trace	.13	1.5	—	—	30	13	64	1.6	28	348	29
Grapefruit:													
Raw:													
Pink, red, white, all varieties	80	.04	.02	.2	21	2.8	38	16	16	.4	1	135	12
Canned, segments, solids and liquid, water pack, without added sweetener	10	.03	.02	.2	20	—	30	13	14	.3	4	144	11
Grapes:													
Raw:													
American type (slip skin) as Concord, Delaware, Niagara, Catawba	100	.05	.03	.3	85	5.2	4	16	12	.4	3	158	13

Amounts in 100 grams, edible portion

FRUITS

FOOD AND DESCRIPTION	Vitamins								Minerals					
	VIT. A	THIAMIN B₁	RIBOFLAVIN B₂	NIACIN	PYRIDOXINE B₆	FOLIC ACID	VIT. C	CALCIUM	PHOSPHORUS	IRON	SODIUM	POTASSIUM	MAGNESIUM
	Internat'l Units	Mg.	Mg.	Mg.	Micro.	Micro.	Mg.	Mg.	Mg.	Mg.	Mg.	Mg.	Mg.
European type (adherent skin) as Malaga, Muscat, Thompson seedless	100	.05	.03	.3	85	5.2	4	12	20	.4	3	173	6
Canned, solids and liquid, water pack, without added sweetener	70	.04	.01	.2	—	—	2	8	13	.3	4	110	—
Guavas, whole raw, common	280	.05	.05	1.2	—	—	242	23	42	.9	4	289	13
Honeydew melon. See Muskmelons													
Kumquats, raw	20	.06	.03	.2	—	—	43	63	23	.4	7	236	—
Lemons, raw:													
Peeled fruit	20	.04	.02	.1	—	—	53	26	16	.6	2	138	—
Fruit, including peel	30	.05	.04	.2	—	7.4	77	61	15	.7	3	145	—
Limes, acid type, raw	10	.03	.02	.2	—	4.6	37	33	18	.6	2	102	—
Loganberries:													
Raw	200	.03	.04	.4	—	—	24	35	17	1.2	1	170	25
Canned, solids and liquid, waterpack, without added sweetener	140	.01	.02	.2	—	—	8	24	11	.8	1	115	—
Juice pack, unsweetened	150	.02	.03	.3	—	—	12	27	15	1.2	1	170	—
Mangos, raw	4,800	.05	.05	1.1	—	—	35	10	13	.4	7	189	18
Muskmelons: Raw:													
Cantaloupes, other netted varieties	3,400	.04	.03	.6	36	6.8	33	14	16	.4	12	251	16
Casaba (Golden Beauty)	30	.04	.03	.6	—	—	13	14	16	.4	12	251	—
Honeydew	40	.04	.03	.6	56	5.0	23	14	16	.4	12	251	—

continued

Amounts in 100 grams, edible portion

FRUITS FOOD AND DESCRIPTION	VIT. A	Vitamins							Minerals					
		THIAMIN B_1	RIBOFLAVIN B_2	NIACIN	PYRIDOXINE B_6	FOLIC ACID	VIT. C		CALCIUM	PHOSPHORUS	IRON	SODIUM	POTASSIUM	MAGNESIUM
	Internat'l Units	Mg.	Mg.	Mg.	Micro.	Micro.	Mg.		Mg.	Mg.	Mg.	Mg.	Mg.	Mg.
Nectarines, raw	1,650	—	—	—	17	20.0	13		4	24	.5	6	294	13
Olives, pickled, canned or bottled:														
Green	300	—	—	—	—	—	—		61	17	1.6	2,400	55	22
Ripe:														
Ascolano (extra large, mammoth, giant, jumbo)	60	trace	trace	—	14	1.0	—		84	16	1.6	813	34	—
Manzanilla (small, medium, large, extra large)	60	trace	trace	—	14	1.0	—		84	16	1.6	813	34	—
Mission (small, medium, large, extra large)	70	trace	trace	—	14	1.0	—		106	17	1.7	750	27	—
Sevillano (giant, jumbo, colossal, super-colossal)	60	trace	trace	—	14	1.0	—		74	20	1.6	828	44	—
Oranges, raw peeled fruit:														
All commercial varieties	200	.10	.04	.4	31	5.1	50		41	20	.4	1	200	11
Canned, unsweetened	200	.07	.02	.3	—	—	40		10	18	.4	1	199	—
Papaws, common North American type, raw	—	—	—	—	35	2.0	—		—	—	—	—	—	—
Papayas, raw	1,750	.04	.04	.3	—	—	56		203	63	6.2	45	727	—
Peaches:														
Raw	1,330	.02	.05	1.0	20	4.0	7		9	19	.5	1	202	10
Canned, solids and liquid, water pack, without added sweetener	450	.01	.03	.6	—	—	3		4	13	.3	2	137	—
Juice pack, unsweetened	670	.01	.04	.9	23	0.5	4		6	19	.5	2	205	—
Dried:														
Uncooked	3,900	.01	.19	5.3	100	5.0	18		45	117	6.0	16	950	48
Cooked without added sugar	1,220	trace	.06	1.5	—	—	2		15	37	1.9	5	297	15
Pears:														
Raw including skin	20	.02	.04	.1	17	2.0	4		8	11	.3	2	130	7

Amounts in 100 grams, edible portion

FRUITS

FOOD AND DESCRIPTION	VIT. A	Vitamins							Minerals					
		THIAMIN B_1	RIBOFLAVIN B_2	NIACIN	PYRIDOXINE B_6	FOLIC ACID	VIT. C	CALCIUM	PHOSPHORUS	IRON	SODIUM	POTASSIUM	MAGNESIUM
	Internat'l Units	Mg.	Mg.	Mg.	Micro.	Micro.	Mg.	Mg.	Mg.	Mg.	Mg.	Mg.	Mg.
Canned, solids and liquid, waterpack, without added sweetener	trace	.01	.02	.1	—	—	1	5	7	.2	1	88	5
Juice pack, unsweetened	trace	.02	.03	.1	14	—	2	8	11	.3	1	130	5
Dried:													
Uncooked	70	.01	.18	.6	—	—	7	35	48	1.3	7	573	31
Cooked, without added sugar	30	trace	.08	.3	—	—	2	16	23	.6	3	269	15
Persimmons, raw:													
Japanese or kaki	2,710	.03	.02	.1	—	—	11	6	26	.3	6	174	8
Native	—	—	—	—	—	—	66	27	26	2.5	1	310	—
Pineapple:													
Raw	70	.09	.03	.2	75	6.0	17	17	8	.5	1	146	13
Canned, solids and liquid, waterpack, without added sweetener	50	.08	.02	.2	—	—	7	12	5	.3	1	99	8
Juice pack, unsweetened	60	.10	.03	.3	71	0.8	10	16	8	.4	1	147	8
Plantain (baking banana) raw	100-1,000	.06	.04	.6	—	—	14	7	30	.7	5	385	—
Plums:													
Raw:													
Damson	300	.08	.03	.5	52	2.0	—	18	17	.5	2	299	9
Japanese and hybrid	250	.03	.03	.5	52	2.0	6	12	18	.5	1	170	—
Prune-type	300	.03	.03	.5	52	2.0	4	12	18	.5	1	170	—
Canned, solids and liquid, without added sweetener:													
Greengage	160	.01	.02	.3	—	—	2	9	13	.2	1	82	—
Purple (Italian Prunes)	1,250	.02	.02	.4	27	1.0	2	9	10	1.0	2	148	—
Pomegranate pulp, raw	trace	.03	.03	.3	—	—	4	3	8	.3	3	259	—
Prunes:													
Dried:													
Uncooked	1,600	.09	.17	1.6	240	5.0	3	51	79	3.9	8	694	40
Cooked, without added sugar	750	.03	.07	.7	—	—	1	24	37	1.8	4	327	20

continued

Amounts in 100 grams, edible portion

FRUITS

FOOD AND DESCRIPTION	VIT. A	Vitamins							Minerals					
		THIAMIN B_1	RIBOFLAVIN B_2	NIACIN	PYRIDOXINE B_6	FOLIC ACID	VIT. C	CALCIUM	PHOSPHORUS	IRON	SODIUM	POTASSIUM	MAGNESIUM
	Internat'l Units	Mg.	Mg.	Mg.	Micro.	Micro.	Mg.	Mg.	Mg.	Mg.	Mg.	Mg.	Mg.
Quinces, raw	40	.02	.03	.2	—	—	15	11	17	.7	4	194	—
Raisins, natural (unbleached), uncooked	20	.11	.08	.5	327	10.0	1	62	101	3.5	27	763	35
Raspberries:													
Raw:													
Black	trace	.03	.09	.9	—	—	18	30	22	.9	1	199	30
Red	130	.03	.09	.9	60	5.0	25	22	22	.9	1	168	20
Canned, solids and liquid, water pack without added sweetener:													
Black	trace	.01	.04	.5	—	—	6	20	15	.6	1	135	—
Red	90	.01	.04	.5	—	—	9	15	15	.6	1	114	13
Sapodilla, raw	60	trace	.02	.2	—	—	14	21	12	.8	12	193	—
Strawberries:													
Raw	60	.03	.07	.6	55	9.0	59	21	21	1.0	1	164	12
Canned, solids and liquid, water pack, without added sweetener	40	.01	.03	.4	—	—	20	14	14	.7	1	111	—
Tangerines, raw (Dancy variety)	420	.06	.02	.1	67	1.0	31	40	18	.4	2	126	—
Watermelon, raw	590	.03	.03	.2	33	0.6	7	7	10	.5	1	100	8

COMPOSITION OF FRUITS (continued)

Amounts in 100 grams, edible portion

FRUITS

FOOD AND DESCRIPTION	WATER	FOOD ENERGY	PRO-TEIN	CARBOHYDRATE TOTAL	CARBOHYDRATE FIBER	ASH	TOTAL FAT	MISCELLANEOUS AND COMMENTS
	percent	calories	grams	grams	grams	grams	grams	
Apples:								
Raw, common varieties:								Contains digestive agent pectin. Contains less sugar than most other fruits.
Freshly harvested and stored, not pared	89.4	58	.2	14.5	1.0	.3	.6	
Freshly harvested, not pared	84.4	56	.2	14.1	1.0	.3	.6	
Stored, not pared	83.9	60	.2	14.8	1.0	.3	.7	
Dried:								
Uncooked	24.0	275	1.0	71.8	3.1	1.6	1.6	
Cooked, without added sugar	78.4	78	.3	20.3	.9	.5	.5	
Apricots:								High in Vitamin A and potassium, particularly in dried state; seed in pit contains Laetrile.
Raw	85.3	51	1.0	12.8	.6	.7	.2	
Canned, solids and liquid, water pack, without added sweetener	89.1	38	.7	9.6	.4	.5	.1	
Dried:								
Uncooked	25.0	260	5.0	66.5	3.0	3.0	.5	
Cooked, without added sugar	76.6	85	1.6	21.6	1.0	1.0	.2	
Avocados, raw, all commercial varieties	74.0	167	2.1	6.3	1.6	1.2	16.4	Highest in unsaturated fat (except for olives), high in protein.
Bananas:								High in digestible sugar when eaten ripe; highest in B_6 of all fruits; high in zinc.
Raw:								
Common	75.7	85	1.1	22.2	.5	.8	.2	
Red	74.4	90	1.2	23.4	.4	.8	.2	
Blackberries:								
Raw	84.5	58	1.2	12.9	4.1	.5	.9	
Canned, solids and liquid, juice pack	85.8	54	.8	12.1	2.7	.5	.8	
Blueberries:								
Raw	83.2	62	.7	15.3	1.5	.3	.5	
Canned, solids and liquid, water pack, without added sweetener	89.3	39	.5	9.8	1.0	.2	.2	

continued

253

Amounts in 100 grams, edible portion

FRUITS

FOOD AND DESCRIPTION	WATER	FOOD ENERGY	PRO-TEIN	CARBOHYDRATE TOTAL	CARBOHYDRATE FIBER	ASH	TOTAL FAT	MISCELLANEOUS AND COMMENTS
	percent	calories	grams	grams	grams	grams	grams	
Boysenberries: Canned, solids and liquid, water pack, without added sweetener	89.8	36	.7	9.1	1.9	.3	.1	
Breadfruit, raw	70.8	103	1.7	26.2	1.2	1.0	.3	
Cantaloupe, see Muskmelons								
Carambola, raw	90.4	35	.7	8.0	.9	.4	.5	
Cherimoya, raw	73.5	94	1.3	24.0	2.2	.8	.4	Avoid drinking water when eating raw.
Cherries: Raw: Sour red	83.7	58	1.2	14.3	.2	.5	.3	
Sweet	80.4	70	1.3	17.4	.4	.6	.3	
Canned, solids and liquid, water pack, without added sweetener: Sour red	88.0	43	.8	10.7	.1	.3	.2	
Sweet	86.6	48	.9	11.9	.3	.4	.2	
Coconut meat: Fresh	50.9	346	3.5	9.4	4.0	.9	35.3	Only fruit high in saturated fats in raw state.
Dried, unsweetened	3.5	662	7.2	23.0	3.9	1.4	64.9	
Coconut Milk (liquid extracted from mixture of grated coconut meat and water)	65.7	252	3.2	5.2	—	1.0	24.9	
Crabapples, raw	81.1	68	.4	17.8	.6	.4	.3	
Cranberries, raw	87.9	46	.4	10.8	1.4	.2	.7	

Amounts in 100 grams, edible portion

FRUITS

FOOD AND DESCRIPTION	WATER	FOOD ENERGY	PRO-TEIN	CARBOHYDRATE TOTAL	CARBOHYDRATE FIBER	ASH	TOTAL FAT	MISCELLANEOUS AND COMMENTS
	percent	calories	grams	grams	grams	grams	grams	
Currants, raw:								
Black European	84.2	54	1.7	13.1	2.4	.9	.1	
Red and white	85.7	50	1.4	12.1	3.4	.6	.2	
Dates, domestic, natural and dried	22.5	247	2.2	72.9	2.3	1.9	.5	
Figs:								
Raw	77.5	80	1.2	20.3	1.2	.7	.3	
Canned, solids and liquid, water pack, without added sweetener	86.6	48	.5	12.4	.7	.3	.2	
Dried, uncooked	23.0	274	4.3	69.1	5.6	2.3	1.3	
Gooseberries:								
Raw	88.9	39	.8	9.7	1.9	.4	.2	
Canned, solids and liquid, water pack, without added sweetener	92.5	26	.5	6.6	1.3	.3	.1	
Granadilla, purple (passion fruit) pulp and seeds, raw	75.1	90	2.2	21.2	—	.8	.7	
Grapefruit:								
Raw: Pink, red, white, all varieties	88.4	41	.5	10.6	.2	.4	.1	
Canned, segments, solids and liquid, water pack without added sweetener	91.3	30	.6	7.6	.2	.4	.1	
Grapes:								
Raw: American type (slip skin) as Concord, Delaware, Niagara, Catawba	81.6	69	1.3	15.7	.6	.4	1.0	Grapes and juice are used as body cleanser and purifier.

continued

255

Amounts in 100 grams, edible portion

FRUITS

FOOD AND DESCRIPTION	WATER	FOOD ENERGY	PRO-TEIN	CARBOHYDRATE TOTAL	FIBER	ASH	TOTAL FAT	MISCELLANEOUS AND COMMENTS
	percent	calories	grams	grams	grams	grams	grams	
Grapes, continued								
European type (adherent skin) as Malaga, Muscat, Thompson seedless	81.4	67	.6	17.3	.5	.4	.3	
Canned, solids and liquid, water pack, without added sweetener	85.5	51	.5	13.6	.2	.3	.1	
Guavas, whole, raw, common	83.0	62	.8	15.0	5.6	.6	.6	High in Vitamin C.
Honeydew melon, see Muskmelons								
Kumquats, raw	81.3	65	.9	17.1	3.7	.6	.1	
Lemons, raw:								Used as purifier, antiseptic and tenderizer; low in sugar.
Peeled fruit	91.9	27	1.1	8.2	.4	.3	.3	
Fruit, including peel	87.4	20	1.2	10.7	—	.4	.3	
Limes, acid type, raw	89.3	28	.7	9.9	.5	.3	.2	
Loganberries:								
Raw	83.0	62	1.0	14.9	3.0	.5	.6	
Canned, solids and liquid, waterpack, without added sweetener	89.2	40	.7	9.4	2.0	.3	.4	
Juice pack, unsweetened	85.7	54	.7	12.7	2.1	.4	.5	
Mangos, raw	81.7	66	.7	16.8	.9	.4	.4	High in Vitamin A.
Muskmelons:								Over 90% water; lowest in sugar of all fruits.
Raw:								
Cantaloupes, other netted varieties	91.2	30	.7	7.5	.3	.5	.1	
Casaba, (Golden Beauty)	91.5	27	1.2	6.5	.5	.8	trace	
Honeydew	90.6	33	.8	7.7	.6	.6	.3	

Amounts in 100 grams, edible portion

FRUITS

FOOD AND DESCRIPTION	WATER	FOOD ENERGY	PRO-TEIN	CARBOHYDRATE TOTAL	CARBOHYDRATE FIBER	ASH	TOTAL FAT	MISCELLANEOUS AND COMMENTS
	percent	calories	grams	grams	grams	grams	grams	
Nectarines, raw	81.8	64	.6	17.1	.4	.5	trace	
Olives, pickled, canned or bottled:								Ripe olives highest in unsaturated fat of all fruits.
Green	78.2	116	1.4	1.3	1.3	6.4	12.7	
Ripe:								
Ascolano (extra large, mammoth, giant, jumbo)	80.0	129	1.1	2.6	1.4	2.5	13.8	
Manzanilla (small, medium, large, extra large)	80.0	129	1.1	2.6	1.4	2.5	13.8	
Mission (small, medium, large, extra large)	73.0	184	1.2	3.2	1.5	2.5	20.1	
Sevillano (giant, jumbo, colossal, super-colossal)	84.4	93	1.1	2.7	1.2	2.3	9.5	
Oranges, raw, peeled fruit:								
All commercial varieties	86.0	49	1.0	12.2	.5	.6	.2	
Canned, unsweetened	87.4	48	.8	11.2	.1	.4	.2	
Papaws, common North American type, raw	76.6	85	5.2	16.8	—	.5	.9	
Papayas, raw	88.7	39	.6	10.0	.9	.6	.1	Contains papain; is a digestant and tenderizer.
Peaches:								
Raw	89.1	38	.6	9.7	.6	.5	.1	
Canned, solids and liquid, water pack, without added sweetener	91.1	31	.4	8.1	.4	.3	.1	
Juice pack, unsweetened	87.2	45	.6	11.6	.4	.5	.1	
Dried:								
Uncooked	25.0	262	3.1	68.3	3.1	2.9	.7	
Cooked without added sugar	76.5	82	1.0	21.4	1.0	.9	.2	
Pears:								
Raw, including skin	83.2	61	.7	15.3	1.4	.4	.4	

continued

Amounts in 100 grams, edible portion

FRUITS

FOOD AND DESCRIPTION	WATER	FOOD ENERGY	PRO-TEIN	CARBOHYDRATE TOTAL	CARBOHYDRATE FIBER	ASH	TOTAL FAT	MISCELLANEOUS AND COMMENTS
	percent	calories	grams	grams	grams	grams	grams	
Pears, continued								
Canned, solids and liquid, waterpack, without added sweetener	91.1	32	.2	8.3	.7	.2	.2	
Juice pack, unsweetened	87.3	46	.3	11.8	.8	.3	.3	
Dried:								
Uncooked	26.0	268	3.1	67.3	6.2	1.8	1.8	
Cooked, without added sugar	65.2	126	1.5	31.7	2.9	.8	.8	
Persimmons, raw:								Good for digestion.
Japanese or kaki	78.6	77	.7	19.7	1.6	.6	.4	
Native	64.4	127	.8	33.5	1.5	.5	.4	
Pineapple:								See Papaya.
Raw	85.3	52	.4	13.7	.04	.04	.2	
Canned, solids and liquid, waterpack, without added sweetener	89.1	39	.3	10.2	.3	.3	.1	
Juice pack, unsweetened	84.0	58	.4	15.1	.3	.4	.1	
Plantain (baking banana), raw	66.4	119	1.1	31.2	.4	.9	.4	Highest in starch of all fruits.
Plums:								
Raw:								
Damson	81.1	66	.5	17.8	.4	.6	trace	
Japanese and hybrid	86.6	48	.5	12.3	.6	.4	.2	
Prune-type	78.7	75	.8	19.7	.4	.6	.2	
Canned, solids and liquid, without added sweetener:								
Greengage	90.6	33	.4	8.6	.2	.3	.1	
Purple (Italian Prunes)	86.8	46	.4	11.9	.3	.7	.2	
Pomegranate pulp, raw	82.3	63	.5	16.4	.2	.5	.3	Juice is a good tonic.
Prunes:								Especially good as a laxative.
Dried:								
Uncooked	28.0	255	2.1	67.4	1.6	1.9	.6	
Cooked, without added sugar	66.4	119	1.0	31.4	.8	.9	.3	

Amounts in 100 grams, edible portion

FRUITS

FOOD AND DESCRIPTION	WATER	FOOD ENERGY	PRO-TEIN	CARBOHYDRATE TOTAL	CARBOHYDRATE FIBER	ASH	TOTAL FAT	MISCELLANEOUS AND COMMENTS
	percent	calories	grams	grams	grams	grams	grams	
Quinces, raw	83.8	57	.4	15.3	1.7	.4	.1	
Raisins, natural (unbleached), uncooked	18.0	289	2.5	77.4	.9	1.9	.2	
Raspberries:								
Raw:								
Black	80.4	73	1.5	15.7	5.1	.6	1.4	
Red	84.2	57	1.2	13.6	3.0	.5	.5	
Canned, solids and liquid, water pack, without added sweetener:								
Black	86.7	51	1.1	10.7	3.3	.4	1.1	
Red	90.1	35	.7	8.8	2.6	.3	.1	
Sapodilla, raw	76.1	89	.5	21.8	1.4	.5	1.1	
Strawberries:								
Raw	89.9	37	.7	8.4	1.3	.5	.5	
Canned, solids and liquid, water pack, without added sweetener	93.7	22	.4	5.6	.6	.2	.1	
Tangerines, raw (Dancy variety)	87.0	46	.8	11.6	.5	.4	.2	
Watermelon, raw	92.6	26	.5	6.4	.3	.3	.2	See Muskmelons; also used as body cleanser; high in inositol.

FRUITS (For explanation of terms, see p. 226)

Yin and Yang

Yang △	Yin ▽
More Yang △△	More Yin ▽▽
	Very Yin ▽▽▽

Apple △△▽▽
Banana ▽▽▽
Cherry △
Fig ▽▽
Grapefruit ▽▽▽
Lime ▽▽▽
Mango ▽▽▽
Melon ▽▽▽
Orange ▽▽▽
Papaya ▽▽▽
Peach ▽
Pineapple ▽▽▽

Acid and Alkaline Content of Fruits

Alkaline Ash		Acid Ash
Apple	Lime	Cranberries
Apricots, raw	Loganberries	Plums
Apricots, dried	Mango	Prunes
Banana	Nectarines	
Blackberries, raw	Orange	
Blueberries, raw	Peach, raw	
Cantaloupe	Pear, raw	
Cherries, fresh	Persimmon	
Currants, fresh	Pineapple, raw	
Dates, dried	Raisins	
Figs, dried	Raspberries, black and red	
Gooseberries	Strawberries	
Grapefruit	Tangerine	
Grapes	Watermelon	
Lemon		

Nut and Seed Glossary

Most nuts and seeds can be served separately (raw or roasted) or in combination with other food as snacks and appetizers. Some interesting combinations are: olives stuffed with almonds; or a spread made from cream cheese, minced clams, curry powder, and chopped Brazil nuts. Macadamia nuts are usually served with drinks. Seeds can be combined with other seeds, nuts, or dried fruit.

Nuts are used extensively in baking cakes, cookies, and breads, and in candy and ice cream. Those most commonly used are almonds (also as a paste for marzipan; and sweet or bitter almond extract), pecans, walnuts, pistachios, and hazelnuts. Most nuts combine well with fruits, raw or cooked.

Cooked nuts and seeds often make an exciting addition to main dishes, rice, or vegetables. Almonds are popular in many Middle Eastern and Mediterranean dishes, also in curries and pilafs. Peanuts can be cooked in a variety of ways: as a vegetable or in soup. Chestnuts often appear in recipes for stuffings but also are good in polenta and pudding, or as a confection (*marrons glacés*). Pignoli are good in *dolmas* (stuffed grape or cabbage leaves).

Nut butters (either homemade or purchased) and nut milks (made in a blender) most often are made from peanuts, cashews, and almonds. Some seeds (like sesame) can be made into pastes; some (like alfalfa) can be sprouted in the same manner as legumes; some (like cumin) are used as seasoning.

Buying and Storing Nuts can be purchased in shells or shelled, roasted, and salted. Raw, unprocessed nuts and seeds can be found at natural-food stores. Most nuts last longer in their shells (walnuts, up to six months; peanuts, up to two years). Opened cans and jars of shelled nuts should be stored in the refrigerator to avoid the nuts' becoming rancid. It is possible to freeze shelled nuts; make sure they're in an airtight container. Some sample equivalent quantities are: 1 pound of almonds with shells equals 1 cup shelled; 1 pound shelled and steamed chestnuts equals 1 cup pureed; 1 pound (approximately 40) Brazil nuts equals 2 cups shelled and chopped; 1 pound pecans with shells equals approximately 1½ cups; and 1 pound walnuts with shells equals 2 cups.

Shelling Most nuts are easily shelled with a nutcracker, but some that cling tightly to the shell can be difficult. To shell pecans easily, moisten the nuts by spreading between damp towels or toweling for a few hours. Afterward, crack nuts at both ends (not sides) and, if necessary, restore crispness by placing in a slow oven for a few minutes. Brazil nuts, also difficult to shell, may be placed in a pan covered with water and heated for about fifteen minutes, then cracked lengthwise. When roasting chestnuts, slash them with a sharp knife on the round side before placing in oven and shake occasionally.

Roasting Some nutritionists contend that it is better to dry nuts at moderate temperatures, just to make them crisper, than to roast them, because high temperatures destroy the B-complex vitamins. If you want to roast shelled nuts, mix a teaspoon of oil with 1 cup of nuts and spread on a shallow pan or cookie sheet. Heat in a 350° F. oven for 5 to 10 minutes, or until lightly browned. Peanuts may be roasted in the shell without oil for a period of 15 to 20 minutes.

Blanching Set nuts in boiling water until skin wrinkles; then quickly rinse in cold water and remove skin. Peanuts can be heated slightly and then pressed gently with a rolling pin to remove inner skins. The inner skins of nuts contain vitamins, so it is best to be moderate about blanching.

Grinding can be done in a blender or small electric grinder. Nut butters can be made this way without adding liquid, providing the nuts are crisp.

Chopping is best done with a chopper or large knife in a wooden bowl.

Slicing is easier if nuts are slightly warm.

Baking If nuts are warmed before adding to batters, they will not sink to the bottom.

Nuts are also used in the manufacture of non-food items such as cosmetics, linoleums, lubricating oils, pharmaceutical preparations, polish, soap, and varnishes. Excellent cooking and salad oils are extracted from nuts and seeds.

COMPOSITION OF NUTS

NUTS AND SEEDS

Amounts in 100 grams, edible portion

FOOD AND DESCRIPTION	VIT. A	Vitamins							Minerals					
		THIA-MIN B_1	RIBO-FLAVIN B_2	NIACIN	PYRI-DOXINE B_6	FOLIC ACID	VIT. C	CAL-CIUM	PHOS-PHORUS	IRON	SODIUM	POTAS-SIUM	MAG-NESIUM
	Internat'l Units	Mg.	Mg.	Mg.	Micro.	Micro.	Mg.	Mg.	Mg.	Mg.	Mg.	Mg.	Mg.
Almonds, dried	0	.24	.92	3.5	100	45.0	trace	234	504	4.7	4	773	270
Beechnuts	—	—	—	—	—	—	—	—	—	—	—	—	—
Brazil nuts	trace	.96	.12	1.6	170	45.0	—	186	693	3.4	1	715	225
Butternuts	—	—	—	—	—	—	—	—	—	6.8	—	—	—
Cashew nuts	100	.43	.25	1.8	0	—	—	—	—	.4	60	330	267
Chestnuts:													
Fresh	—	.22	.22	.6	—	—	—	27	88	1.7	6	454	41
Dried	—	.32	.38	1.2	—	—	—	52	162	3.3	12	875	—
Filberts (Hazelnuts)	—	.46	—	.9	—	66.6	trace	209	337	3.4	2	704	184
Hickory nuts	—	—	—	—	—	—	—	trace	360	2.4	—	—	160
Lychees:													
Raw	—	—	.05	—	—	—	42	8	42	.4	3	170	—
Dried	—	—	—	—	—	—	—	33	181	1.7	3	1,100	—
Macadamia nuts	0	.34	.11	1.3	—	—	0	48	161	2.0	—	264	—
Peanuts:													
Raw with skins	—	1.14	.13	17.2	—	—	0	69	401	2.1	5	674	206
Raw without skins	0	.99	.13	15.8	—	—	0	59	409	2.0	5	674	—
Roasted with skins	—	.32	.13	17.1	300	56.5	0	72	407	2.2	5	701	175
Pecans	130	.86	.13	.9	183	27.0	2	73	289	2.4	trace	603	142

263

Amounts in 100 grams, edible portion

NUTS AND SEEDS

FOOD AND DESCRIPTION	VIT. A	Vitamins						Minerals					
		THIAMIN B_1	RIBOFLAVIN B_2	NIACIN	PYRIDOXINE B_6	FOLIC ACID	VIT. C	CALCIUM	PHOSPHORUS	IRON	SODIUM	POTASSIUM	MAGNESIUM
	Internat'l Units	Mg.	Mg.	Mg.	Micro.	Micro.	Mg.	Mg.	Mg.	Mg.	Mg.	Mg.	Mg.
Pine nuts:													
Pignolias	—	.62	—	—	—	—	—	—	—	—	—	—	—
Piñon	30	1.28	.23	4.5	—	—	trace	12	604	5.2	—	—	—
Pistachio nuts	230	.67	—	1.4	—	—	0	131	500	7.3	—	972	158
Pumpkin and squash seed kernels; dry	70	.24	.19	2.4	—	—	—	51	1,144	11.2	—	—	—
Safflower seed kernels, dry	—	—	—	—	110	.1	—	—	—	—	—	—	—
Sesame seeds, dry, whole	30	.98	.24	5.4	—	—	0	1,160	616	10.5	60	725	181
Sunflower seed kernels, dry	50	1.96	.23	5.4	—	—	—	120	837	7.1	30	920	38
Walnuts:													
Black	300	.22	.11	.7	—	—	—	trace	570	6.0	3	460	190
Persian or English	30	.33	.13	.9	960	77.0	2	99	380	3.1	2	450	130

NOTE:

All dried fruits are high in calories and generally high in potassium.
All citrus are high in Vitamin C; inside of skins are high in bioflavonoids.
All nuts and seeds are very high in protein, fat and calories.

COMPOSITION OF NUTS (continued)

Amounts in 100 grams, edible portion

NUTS AND SEEDS

FOOD AND DESCRIPTION	WATER	FOOD ENERGY	PROTEIN	CARBOHYDRATE TOTAL	CARBOHYDRATE FIBER	ASH	TOTAL FAT	MISCELLANEOUS AND COMMENTS
	percent	calories	grams	grams	grams	grams	grams	
Almonds, dried	4.7	589	18.6	19.5	2.6	3.0	54.2	
Beechnuts	6.6	568	19.4	20.3	3.7	3.7	50.0	
Brazil nuts	4.6	654	14.3	10.9	3.1	3.3	66.9	Highest in fat and calories of all nuts.
Butternuts	3.8	629	23.7	8.4	—	2.9	61.2	
Cashew nuts	5.2	561	17.2	29.3	1.4	2.6	45.7	
Chestnuts:								
Fresh	52.5	194	2.9	42.1	1.1	1.0	1.5	
Dried	8.4	377	6.7	78.6	2.5	2.2	4.1	
Filberts (Hazelnuts)	5.8	634	12.6	16.7	3.0	2.5	62.4	
Hickory nuts	3.3	673	13.2	12.8	1.9	2.0	68.7	
Lychees:								
Raw	81.9	64	.9	16.4	.3	.5	.3	
Dried	22.3	277	3.8	70.7	1.4	2.0	1.2	
Macadamia nuts	3.0	691	7.8	15.9	2.5	1.7	71.6	
Peanuts:								
Raw, with skins	5.6	564	26.0	18.6	2.4	2.3	47.5	
Raw, without skins	5.4	568	26.3	17.6	1.9	2.3	48.4	
Roasted with skins	1.8	582	26.2	20.6	2.7	2.7	48.7	
Pecans	3.4	687	9.2	14.6	2.3	1.6	71.2	
Pine nuts:								
Pignolias	5.6	552	31.1	11.6	.9	4.3	47.4	
Piñon	3.1	635	13.0	20.5	1.1	2.9	60.5	

265

Amounts in 100 grams, edible portion

NUTS AND SEEDS

FOOD AND DESCRIPTION	WATER	FOOD ENERGY	PRO-TEIN	CARBOHYDRATE TOTAL	FIBER	ASH	TOTAL FAT	MISCELLANEOUS AND COMMENTS
	percent	calories	grams	grams	grams	grams	grams	
Pistachio nuts	5.3	594	19.3	19.0	1.9	2.7	53.7	
Pumpkin and squash seed kernels, dry	4.4	553	29.0	15.0	1.9	4.9	46.7	
Safflower seed kernels, dry	5.0	615	19.1	12.4	—	4.0	59.9	
Sesame seeds, dry, whole	5.4	563	18.6	21.6	6.3	5.3	49.1	
Sunflower seed kernels, dry	4.8	560	24.0	19.9	3.9	4.0	47.3	
Walnuts:								
Black	3.1	628	20.5	14.3	1.7	2.3	59.3	
Persian or English	3.5	651	14.8	15.8	2.1	1.9	64.0	

NUTS (For explanation of terms, see p. 226)

Yin and Yang			
Yang △	Yin ▽		
	More Yin ▽▽		
Almond ▽▽	Chestnut △▽	Peanut ▽▽	
Cashew ▽▽	Hazelnut △▽		

Acid and Alkaline Content of Nuts

Alkaline Ash	Acid Ash
Almonds	Brazil nuts
Chestnuts	Peanuts
Coconut, fresh	Walnuts, English

Recipes

INDIAN-STYLE VEGETABLE APPETIZERS

Serves 4 to 6

- 1 cup unbleached or wholewheat flour
- ½ tablespoon cayenne pepper
- 1 tablespoon ground cumin
- 2 tablespoons turmeric
- 1 teaspoon cinnamon
- 1 teaspoon ground ginger (optional)
- cold water
- 1 potato, sliced thin
- 1 zucchini, sliced thin
- 1 green pepper, seeded and sliced thin
- ghee or oil

1. Combine flour, spices, and enough water to make a thick pancake batter.
2. Dip vegetables in batter and deep-fry in ghee (clarified butter) or oil until golden.
3. Drain and serve immediately.

NOTE

Curry powder may be substituted for spices.

RAW CARROT BALLS

Serves 8 to 10

- 8 ounces cream cheese
- yogurt
- 6 medium-sized carrots, shredded finely
- 3 tablespoons wheat germ
- 1 tablespoon honey
- 1 teaspoon sea salt
- ½ cup parsley, chopped fine
- 1 cup pecans and almonds, ground
- 1 tablespoon sweet cream
- lettuce, celery sticks, olives

1. Mash the cheese with a fork and add enough yogurt to make creamy.
2. Add carrots, sweet cream, wheat germ, honey, salt, and parsley.
3. Form into balls the size of cherry tomatoes and roll in the ground nuts.
4. Arrange in a bowl or plate on a bed of lettuce, garnished with celery sticks and olives.

"TAWNEY" NUTS

      oil
- 1 pound raw peanuts, walnuts, or almonds; or any combination of raw nuts and seeds
- 1½ tablespoons honey
- 1 teaspoon anise seed

1. Place nuts in well-oiled baking pan.
2. Put in slow oven (200° to 250° F.) until roasted (approximately 45 minutes); stir occasionally.
3. Remove from oven and sprinkle honey and anise seed over.
4. Put back in oven for a few minutes, then remove and allow to cool.

SWEET MANGO RELISH

Serves 6 to 8

- 4 green mangoes (to equal 1 quart, chopped)
- 2 large onions
- 6 sweet red peppers
- 2 large hot peppers
- 4 cups honey
- 1 tablespoon salt
- 1 tablespoon white mustard seed
- 1 tablespoon celery seed
- 1 cup vinegar
- 2 cups raisins (optional)

1. Peel green mangoes, cut from seed, and chop by hand or chop coarsely in blender enough to make 1 quart.
2. Chop or grind onions, sweet peppers, and hot peppers.
3. Mix other seasonings, vinegar, and raisins.
4. Combine all ingredients, bring to boil, and boil 10 minutes.
5. Let stand overnight.
6. Next morning, cook until slightly thickened.
7. Pack while boiling hot and seal. Keep refrigerated.

PICKLED DAIKON (Oriental Radish)

Yield: about 1½ quarts

¼ cup sea salt
3 daikon, chopped crossways
½ cup tamari (soy sauce)
¾ cup honey (more if desired)
¼ cup vinegar
2 tablespoons fresh ginger, chopped fine
1 chili pepper (optional)

1. Sprinkle salt over daikon and soak overnight; wash and drain.
2. Boil tamari, honey, vinegar, ginger, and chili pepper for 3 minutes.
3. Add daikon and bring to rapid boil.
4. Cook 1 minute and pack in jars. Cool.
5. Store in refrigerator. Use as a relish.

FRESH TURNIP-KRAUT

Serves 3 to 4

2 to 3 turnips
juice of 1 lemon
1½ tablespoons olive oil
sea salt
ripe olives (optional)

1. Shred turnips.
2. Squeeze lemon juice over turnips.
3. Add the oil and salt to taste, and toss. Chill. May be garnished with ripe olives.

GRAPEFRUIT SALAD DRESSING

Yield: 1 cup

1 grapefruit
1 tablespoon honey
1 teaspoon paprika
1 teaspoon mustard
1 small onion, sliced thinly
½ cup vegetable oil

1. Peel grapefruit and put into blender.
2. Add other ingredients.
3. Blend together and use on vegetable or fruit salads.

DRIED FRUIT SOUP

Serves 6 to 8

1	cup large dried apricots
1	cup large dried pitted prunes
½	cup dried currants
¼	cup seedless raisins
7	cups cold water
	cinnamon stick, approximately 1 inch
4 to 5	whole cloves
⅓	cup lemon juice
½	orange (with pulp and skin), chopped
4	slices unsweetened pineapple, cut into small pieces
1	large apple, cored and chopped
2	tablespoons honey
1½	tablespoons cornstarch
2	tablespoons cold water
¼	teaspoon sea salt
	yogurt or mint (optional)

1. Combine all dried fruits with 3 cups water, cinnamon, cloves, and half of lemon juice.
2. Simmer for 20 minutes over medium heat.
3. Add chopped orange, pineapple, apple, honey, remaining lemon juice, and salt.
4. Add the 4 remaining cups of water and simmer for another 15 minutes.
5. Add dissolved mixture of cornstarch and cold water to fruit and cook until slightly thickened, stirring continuously.
6. May be served hot or cold, topped with yogurt or mint.

COLD MELON SOUP

Serves 4 to 6

- 1 large ripe canteloupe
- ½ teaspoon cinnamon
- 2½ cups orange juice
- 3 tablespoons fresh lime juice
- fresh mint

1. Peel and cut melon into cubes.
2. Mix melon, cinnamon, and ½ cup of orange juice in blender.
3. Combine the remaining orange juice and the lime juice and add to blended mixture.
4. Cover and refrigerate for 1 hour or more. Serve garnished with mint.

COLLEEN'S PUMPKIN BREAD

Makes 4 1-pound loaves

- 4 cups unbleached flour
- 3½ cups honey
- 1 cup oil
- ⅔ cup cold water
- 1 pound pumpkin, cooked and mashed
- 2 teaspoons baking soda
- 1½ teaspoons salt
- 4 eggs, added 1 at a time
- 2 teaspoons cinnamon
- 1 teaspoon nutmeg
- ½ teaspoon allspice
- 1 teaspoon baking powder
- ½ teaspoon ground cloves
- 1 cup almonds or assorted nuts, chopped

1. Combine all ingredients and mix thoroughly.
2. Fill 4 1-pound coffee cans or bread pans.
3. Bake for 60 minutes in a 350° F. oven.

PEANUT STUFFING

Yield: 3 cups

- 2 cups stale bread, moistened and crumbled
- rind of 1 lemon, grated
- 1 cup peanuts, chopped finely
- 2 tablespoons mixed herbs
- ½ cup melted butter or oil
- salt and pepper to taste

1. Combine all ingredients and use as stuffing for fowl, meat, or vegetables.

TOFU (BEAN CURD) SAUCE

Yield: approximately 1 cup

- 1 block fresh tofu
- 3 tablespoons sesame seeds, toasted
- 3½ tablespoons honey
- 2 teaspoons tamari (soy sauce)

1. Mash tofu.
2. Grind sesame seeds.
3. Place tofu, sesame seeds, honey, and tamari in blender and blend until smooth.
4. Spoon over any cooked vegetable.

SAUTEED GREEN BEANS

Serves 4 to 5

- 2 large garlic cloves, minced
- oil
- ½ cup hiziki (black seaweed), soaked for approximately 15 minutes
- 1 pound fresh green beans, whole
- 4 tablespoons miso (soybean paste)
- ¾ cup cold water
- 2 teaspoons tamari (soy sauce)

1. Fry garlic in oil until golden.
2. Sauté hiziki for 10 minutes.
3. Add green beans and sauté until tender yet crunchy and slightly browned.
4. Add miso mixed with cold water and tamari.
5. Stir and cook for a few minutes. Serve hot.

CORN-ZUCCHINI PUDDING

Serves 2 to 3

- 1 onion
- ½ green pepper
- 1 clove garlic
- 2 cups zucchini, sliced
- 1 cup raw corn
- 3 eggs

1. Grind onion, green pepper, and garlic.
2. Add zucchini and corn.
3. Separate eggs and mix yolks with vegetables.
4. Beat egg whites until stiff; gently fold into mixture.
5. Bake about 20 minutes in a 350° F. oven.
6. Serve hot.

SAUTEED BEAN SPROUTS WITH GINGER

Serves 2 to 4

    oil
1 pound fresh soybean sprouts
2 tablespoons tamari (soy sauce)
½ teaspoon fresh ginger, minced
1 tablespoon cider or rice vinegar (optional)

1. Heat oil and sauté sprouts, tamari, and ginger.
2. Toss gently for about 5 or 6 minutes, until the sprouts are wilted but not overcooked.
3. Add vinegar if a tart taste is desired.

RAW APPLE SAUCE

Serves 3 to 4

3 to 4 crisp apples
1 tablespoon lemon juice
¼ cup honey
¼ to ½ cup apple juice

1. Wash, core, and dice apples.
2. Place lemon juice, honey, and apple juice in blender and add diced apple until ⅓ full.
3. Blend at medium-fast speed, add remaining diced apple, and continue to blend until smooth.
4. Serve chilled.

AVOCADO-LIME ICE CREAM

Yield: 1 quart

2 soft avocados, peeled, pitted, and cubed
¼ cup lime juice
½ cup honey
1 cup whipped cream

1. Place all ingredients except cream in blender and blend until smooth.
2. Fold in whipped cream.
3. Freeze in freezer tray until firm (4 hours or more).

KANTEN (AGAR-AGAR) DESSERT

Serves 2 to 3

- 1 bar kanten (agar-agar: a gelatine derived from seaweed)
- 1½ cups cold water
- pinch sea salt
- 1½ cups apple juice
- ¾ to 1 cup chopped fresh fruit or whole berries

1. Break and soak kanten in water for 10 minutes.
2. Boil, stirring, until dissolved.
3. Add salt, apple juice, and fresh fruit.
4. Simmer for 10 minutes.
5. Refrigerate until firm.

MIDDLE EASTERN GRAIN PUDDING

Serves 4 to 6

- 4 tablespoons rice or barley, ground fine
- 1 quart milk
- 6 to 8 tablespoons raw sugar or honey
- ½ cup cooked chick peas
- 3 tablespoons rosewater
- ½ cup almonds, chopped fine
- pomegranate kernels
- ground pistachios
- shredded coconut

1. Soak rice or barley in a little cold water for a few minutes and then blend into a smooth paste.
2. Mix milk with sugar or honey and bring to a boil.
3. Add paste gradually and simmer mixture over very low heat, stirring continuously until thick enough to coat spoon. Do not allow milk to burn.
4. Add chick peas and rosewater; stir and simmer for a few minutes.
5. Add almonds and cool mixture slightly; then refrigerate.
6. Serve chilled, topped with pomegranate kernels, ground pistachios, and shredded coconut.

SUGARLESS CHESTNUT PUDDING

Serves 4

1 pound chestnuts
1 tablespoon (1 envelope) unflavored gelatine
4 tablespoons lemon juice
whipped cream (optional)

1. Shell nuts by boiling for a minute before peeling.
2. Replace nuts in boiling water and cook until soft.
3. Put nuts through sieve, food mill, or blender.
4. Mix gelatine and lemon juice and combine with nuts.
5. Rinse a deep dish with cold water and place mixture in moistened dish.
6. Refrigerate until firm.
7. Mixture may be unmolded and served with whipped cream.

CARROT CAKE

- 2 cups unbleached flour
- 2 teaspoons baking powder
- 1½ teaspoons baking soda
- 2 teaspoons ground cinnamon
- ½ teaspoon ground nutmeg
- ½ teaspoon ground ginger
- 1 teaspoon salt
- 2½ cups honey
- 1½ cups oil
- 4 eggs
- 2 cups grated raw carrots
- 1 cup pineapple, cut or mashed
- ½ cup almonds, chopped fine

1. Mix flour, baking powder, baking soda, spices and salt.
2. Blend oil and honey and add eggs 1 at a time.
3. Add remaining ingredients and mix thoroughly.
4. Bake for 30 to 40 minutes in a 350° F. oven.

TOMATO SOUP CAKE

½ cup margarine or oil
¾ cup honey
1 cup tomato soup
2 cups flour
2 teaspoons baking powder
1 teaspoon cinnamon
½ teaspoon cloves
1 teaspoon nutmeg
1 cup raisins
1 cup chopped nuts

1. Blend margarine and honey.
2. Add soup.
3. Sift flour, baking powder, and spices together; stir in raisins and nuts and add to soup mixture.
4. Bake in 350° F. oven for 50 to 60 minutes.

SUPER ENERGIZING TAHINI VEGETABLE FRUIT DRINK

Serves 2

1 stick celery
1 sprig parsley
¼ small raw beet
2 tablespoons tahini (sesame paste)
1 ripe banana
1 teaspoon honey
  cold water
  cold milk

1. Chop celery, parsley, and beet and put into blender with enough water to blend smoothly.
2. Add tahini, banana, and honey.
3. Blend and add more water or milk to make approximately 14 ounces.
4. Pour into 2 large glasses over ice cubes and sip slowly.

CAROB DATE AND NUT ROLL

24 dates, pitted
1 cup pecan or other raw nuts, ground
4 teaspoons carob powder
½ teaspoon sea salt
honey

1. Put dates through food mill or blender.
2. Add carob powder, blending with a fork, salt, and enough honey to make the mixture sticky.
3. Add ¾ of the nuts and form into a thin roll.
4. Coat with additional nuts and refrigerate.
5. When chilled, slice and serve.

PEANUT BUTTER HONEY BALLS

Yield: Approximately 20 balls

1 pound chunky peanut butter (unhydrogenated)
8 tablespoons honey
  non-fat dry milk
  grated coconut or chopped nuts

1. Blend peanut butter and honey together.
2. Gradually add sufficient milk powder to make a fairly stiff consistency.
3. Shape teaspoonfuls into balls and roll in coconut or chopped nuts.

STRAWBERRY MOCK WHIPPED CREAM

½ cup Fearn soya protein powder
4 heaping tablespoons honey
1 cup water
juice of one lemon
½ cup strawberries

Blend all ingredients well, adding additional water if thinner consistency is desired. Refrigerate and serve very cold on cooked fruits or plain cake.

£3.50